D 93

Bibliografische Information der Deutschen Nationalbibliothek

Die Deutsche Nationalbibliothek verzeichnet diese Publikation in der Deutschen Nationalbibliografie; detaillierte bibliografische Daten sind im Internet über http://dnb.d-nb.de abrufbar.

ISBN 978-3-8325-3398-4

Logos Verlag Berlin GmbH
Comeniushof, Gubener Str. 47,
10243 Berlin
Tel.: +49 (0)30 42 85 10 90
Fax: +49 (0)30 42 85 10 92
INTERNET: http://www.logos-verlag.de

Modeling and parameter estimation for heterogeneous cell populations

Von der Fakultät Konstruktions-, Produktions-, und Fahrzeugtechnik und dem Stuttgart Research Centre for Simulation Technology der Universität Stuttgart zur Erlangung der Würde eines Doktors der Ingenieurwissenschaften (Dr.-Ing.) genehmigte Abhandlung

Vorgelegt von

Jan Hasenauer

aus Vaihingen an der Enz

Hauptberichter:	Prof. Dr.-Ing. Frank Allgöwer
Mitberichter:	Prof. Dr. Mustafa Khammash, Ph.D.
	Prof. Dr. rer. nat. Peter Scheurich

Tag der mündlichen Prüfung: 13.02.2013

Institut für Systemtheorie und Regelungstechnik

Universität Stuttgart

2013

Misura tutto ciò che è misurabile e rendi misurabile ciò che non lo è.
(Measure what can be measured, and make measurable what cannot be measured.)

Galileo Galilei

Acknowledgments

My greatest appreciation goes to my advisor Frank Allgöwer. By providing supervision and an unsurpassed academic environment, he made my time at the University of Stuttgart unique. I am very thankful for this opportunity.

I am also very grateful to all former and current members of the Institute of Systems Theory and Automatic Control (IST). Thank your for the coffee breaks, table soccer tournaments, pizza / movie nights, and your friendship. Foremost, I thank Steffen Waldherr for the mentoring during the first years of my Ph.D. studies. For interesting joint projects, insightful discussion, and proof-reading of this thesis, I thank Andrei Kramer, Martin Löhning, Eva-Maria Geissen, Nicole Radde, Marcus Reble, Daniella Schittler, and Patrick Weber. Regarding organizational issues, Beate Spinner and Claudia Vetter have been irreplaceable.

I would like to express my special gratitude appertains to an other member of the IST, my roommate Christian Breindl. Besides computer support and often helpful distraction via rock-paper-scissors games, I deeply appreciate that he is such a dear and helpful friend, who always provides critical and constructive comments.

I am also grateful to Peter Scheurich, who enriched my work through his advice and ideas, as well as Mustafa Khammash for the inspiring time I spent with his group in Santa Barbara and his great enthusiasm. Whilst this thesis does contain only little of the research carried out in collaboration with Peter Scheurich and Mustafa Khammash, they still have strongly influenced my Ph.D. studies.

I would like to thank my collaboration and discussion partners: Christoph Zechner, Jakob Ruess, and Heinz Koeppl from the ETH Zürich; Philipp Rumschinski and Rolf Findeisen from the Otto-von-Guericke-Universität in Magdeburg; and Christine Andres and Tim Hucho from the Max Planck Institute for Molecular Genetics in Berlin. They have broadened my mind by posing interesting / challenging questions.

My research has financially been supported by the German Federal Ministry of Education and Research (BMBF) within the FORSYS-Partner program, and by the German Research Foundation within the Cluster of Excellence in Simulation Technology at the University of Stuttgart. I appreciate my collaborators in both programs, foremost, Malgorzata Doszczak, Karin Erbertseder, Holger Perfahl, Julian Heinrich, Corinna Vehlow, and Daniel Weiskopf.

Last, but by no means least, I would like to thank my family for their love and constant support. Furthermore, I owe sincere gratitude to Christine Andres, who has been a supportive and inspiring partner during the later stages of my thesis.

Jan Hasenauer
Stuttgart, February 2013

Contents

Index of notation

Acronyms

Acronym	**Description**
BIC	Bayesian information criterion
C3(a)	(active) caspase 3
C8(a)	(active) caspase 8
CFDA-SE	carboxyfluorescein diacetate succinimidyl ester
CFSE	carboxyfluorescein succinimidyl ester
CLE	chemical Langevin equation
CME	Chemical Master Equation
DSP	division-structured population
DLSP	division- and label-structured population
EG	exponential growth
FPE	Fokker-Planck equation
I-κB	inhibitor of nuclear factor κB
IMSE	integrated mean square error
LSP	label-structured population
MCMC	Markov chain Monte Carlo
MJP	Markov jump process
NF-κB	nuclear factor κB
ODE	ordinary differential equation
PDE	partial differential equation
RRE	reaction rate equation
SDE	stochastic differential equation
TNF	tumor necrosis factor
TRAIL	TNF related apoptosis inducing ligand
XIAP	X-linked inhibitor of apoptosis protein

Notation

General notation: Throughout this thesis we follow the common statistical notation and denote random variables by capital letters, e.g., X. To describe the probability (density) of a random variable lower-case letters are used. According to this, $p(x)$ denotes the probability density of X.

Mathematics and statistics

Symbol	Description
$\mathbf{1}$	one vector, $\mathbf{1} = [1, \ldots, 1]^{\mathrm{T}}$
$\mathrm{card}(x)$	cardinality of set $\mathcal{X}$
d_x / $d_{\mathcal{X}}$	dimension of the vector x / cardinality of the set $\mathcal{X}$
$\mathrm{erfc}(x)$	error function
$\exp(x)$	exponential function
$\mathbb{E}[X]$	expected value of X
$\log(x)$	natural logarithm
$\log \mathcal{N}(x\|\mu, \sigma^2)$	log-normal distribution
$\mathbb{N}$	natural numbers
$\mathbb{N}_0$	natural numbers including zero
$P(x)\,(p(x)) \in \mathbb{R}_+$	probability mass (density) of x
$P(x, y)\,(p(x, y)) \in \mathbb{R}_+$	joint probability mass (density) of x and y
$P(x\|y)\,(p(x\|y)) \in \mathbb{R}_+$	conditional probability mass (density) of x given y
$\mathbb{R}$	real numbers
$\mathbb{R}_+$	non-negative real numbers
$\mathbb{Z}$	integers
$\{x^k\}_{k=1}^{d_x}$	set containing $x^1, \ldots, x^{d_x}$

Reaction network model

Symbol	Description
$a_j(X_t, \theta)$	propensity function of j-th chemical reaction
$\gamma(X_t, \theta)$	output map $((X_t, \theta) \mapsto Y_t)$
$\mu(X_t, \theta)$	drift function
ν_j	stoichiometry of j-th chemical reaction
R_j	j-th chemical reaction
S_i	i-th chemical species
$\sigma(X_t, \theta)$	diffusion function
θ	vector of model parameters
W_t	Wiener process
X_t $(x(t))$	state of stochastic (deterministic) single cell model at time t
Y_t $(y(t))$	output of stochastic (deterministic) single cell model at time t

Measurement data and parameter estimation

Symbol	Description
$\mathcal{D}^i$	data of a single cell
$\mathcal{D}$	collection of all data
$d_{\mathcal{D}}$	number of measurements
p_a	acceptance probability
$p(\bar{Y}_t \| Y_t)$	probability density of observing the measured output $\bar{Y}_t$ given output Y_t
$\mathbb{P}(\mathcal{D} \| \theta)$	likelihood of $\mathcal{D}$ given θ
$\mathbb{P}(\mathcal{D})$	marginal probability to observe measurement data
$\pi(\theta \| \mathcal{D})$	posterior probability of θ given $\mathcal{D}$
$\pi(\theta)$	prior probability of parameter vector θ
$\mathbb{Q}(\theta^{k+1}, \theta^k)$	transition function
$\mathcal{SD}_j$	data of a single cell snapshot experiment
$\mathcal{SD}_j^{\mathcal{B}}$	binned data of a single cell snapshot experiment
θ^*	maximum a posteriori estimate of parameter vector θ
$\bar{Y}_t^i$	measured, noise corrupted single cell output

Signal transduction in heterogeneous cell populations

Symbol	Description
$c_j^i = \mathbb{P}(\mathcal{D}^i \| \Lambda_j)$	likelihood of observing $\mathcal{D}^i$ in population with $z_0 \sim \Lambda_j(z_0)$
H	bandwidth matrix of kernel
$\mathbb{K}(y \| Y, H)$	kernel for kernel density estimation
$\Lambda_j(z_0)$	j-th ansatz function for parameter and initial state distribution
$\tilde{\mu}(Z_t, t)$	augmented drift function
$p_0(z)(= p_0([x_0^T, \theta^T]^T))$	joint density of parameters and initial conditions
$p_{0,\varphi}(z)$	parametrized joint density of parameters and initial conditions
$[p]_{0,\varphi}^{1-\alpha}(z)$	$100(1-\alpha)\%$ Bayesian confidence interval of $p_{0,\varphi}(z)$
$p(x \| t, p_0)$	density of state x at time t given $p_0(z)$
$p(y \| t, p_0)$ $(\hat{p}(y \| t, p_0))$	(approximated) density of output y at time t given $p_0(z)$
$p(z \| t, p_0)$	density of augmented state z at time t given $p_0(z)$
φ	density parameter vector
$\tilde{\sigma}(Z_t, t)$	augmented diffusion function
$Z_t = [X_t^T, \theta^T]^T$	augmented state of single cell

Proliferation of heterogeneous cell populations

Symbol	Description
$\alpha_i(t)$	rate of cell division in subpopulation i
$\beta_i(t)$	rate of cell death in subpopulation i
c	proportionality constant of label concentration and label fluorescence
γ	label dilution factor
$\bar{H}_t^l$	number of cells measured in histogram bin l at time t
$k(t)$	rate of label degradation
$n(x, i\|t)$	number density of cells with label concentration x and division number i
$N(i\|t)$	number of cells with division number i
$\mathcal{N}(i\|s)$	Laplace transformation of $N(i\|t)$
$n(x\|t)$	number density of cells with label concentration x
$\hat{n}_S(x\|t)$	approximate number density of cells with label concentration x considering only subpopulations 0 to $S-1$
$n(y\|t)$	number density of cells with label induced fluorescence $y = cx$
$n(\bar{y}\|t)$	number density of cells with measured fluorescence $\bar{y} = y + y_a$
$N(t)$	number of cells
$\nu(t, x)$	rate of label concentration change
$p_a(y_a)$	probability distribution of autofluorescence
$p_o(\bar{y})$	probability distribution of outliers in flow cytometry measurements
$p(x\|i, t)$	probability density that a cell in the i-th subpopulation has a label concentration x at time t
Y_a	level of autofluorescence

Abstract

Systems and computational biology exploit mathematical models to understand and predict the dynamics of biological processes. These processes are in general related to information processing and decision making in individual cells, however, most cell systems considered in biological and clinical research are heterogeneous, i.e. the individual members of the cell population differ. Even cells in clonal cell populations may respond differently to the same stimulus. Furthermore, most experimental techniques cannot provide detailed information about the individual cells. This renders the study of cell populations crucial.

In this thesis, we introduce novel tools for the computational modeling of heterogeneous cell systems. The focus is on biologically consistent models for signal transduction processes in and proliferation of cell populations, as well as sophisticated methods for parameter estimation and uncertainty analysis.

For signal transduction processes a new modeling framework, taking into account stochastic and deterministic sources of cell-to-cell variability, is introduced. To unravel the sources and the structure of deterministic variability, we propose a Bayesian estimation approach using commonly available population level data, which may be collected using, e.g., flow cytometers. Our Bayesian approach employs a parametric model of the variability, which enables the decomposition of the estimation problem and ensures efficient computations. The maximum a posteriori estimate can be computed using sequential convex optimizations, while the uncertainties can be assessed by sampling a unimodal posterior distribution.

To complement the work on signal transduction, the proliferation properties of heterogenous cell populations are investigated. Therefore, we introduce a generalized growth model for structured populations which accounts for differences in the label concentration and the division number of individual cells. The analysis of this model using the superposition principle, a decomposition approach, and an invariance concept provides us with partially analytical solutions and an efficient simulation scheme. Combining these two ingredients with a first likelihood function for binned flow cytometry data allows the development of a Bayesian estimation and uncertainty analysis scheme.

In the past, the computational complexity of population models and the lack of consistent likelihood functions prohibited the in-depth assessment of cell system dynamics. In this thesis, we overcome this problem be intertwining modeling and simulation. The resulting models predict the single cell behavior / properties on the basis of population data. This allows for novel insights in population dynamics as well as the model-based characterization of subgroups.

Deutsche Kurzfassung

Problembeschreibung

Ziel der meisten (system-)biologischen Forschungsprojekte ist es, die Prozesse zu verstehen, die intrazellulärer Signalübertragung zu Grunde liegen. Hierfür werden mathematische Modelle verwendet, die verschiedenes Strukturwissen und Daten berücksichtigen. Modelle können genutzt werden, um diese Informationen zusammenzuführen, und erlauben daher eine ganzheitliche Betrachtung.

Die Entwicklung von quantitativen Modellen umfasst drei Schritte: die Herleitung eines geeigneten Modells, die Bestimmung der Modellparameter und die Analyse des Modells (z.B. Validierung). Sowohl das Vorgehen bei der Parameterschätzung als auch die Analyse hängen stark von dem zugrunde liegenden Modell ab. Abhängig von der biologischen Fragestellung kann eine Vielzahl verschiedener Modelle genutzt werden.

In den meisten Systembiologieprojekten werden sogenannte Einzelzellmodelle verwendet, die das Verhalten einer „typischen" Zelle beschreiben. Für diese Einzelzellmodelle existiert eine Vielzahl von Methoden zur Parameterschätzung und Modellanalyse. Unglücklicherweise stammt jedoch ein Großteil der Daten, die in diese Modelle einfließen, aus Experimenten, in denen über die gesamte Zellpopulation gemittelt wird, wie z.B. bei Western Blot Experimenten. Weisen die einzelnen Zellen in der Population große Unterschiede auf, d.h. die Zellpopulation ist stark heterogen, so kann die Anpassung des Einzelzellmodells an die Populationsdaten irreführende Ergebnisse liefern. Insbesondere für Fälle, in denen die Zellen einen Entscheidungsprozess durchlaufen, unter anderem zellulläre Apoptose, Proliferation oder Differenzierung, spiegelt der Mittelwert keine der vorhandenen Subpopulationen wider.

Zur Untersuchung des Verhaltens von heterogenen Zellpopulation werden Populationsmodelle benötigt, die das Verhalten der Gesamtpopulation auf Basis der Einzelzelldynamiken beschreiben. Um eine gute Vorhersagekraft solcher Modelle zu gewährleisten, sind Methoden für die Parameterschätzung und Modellanalyse unerlässlich.

Im Folgenden werden wir aufzeigen, dass die momentan verfügbaren Klassen von Populationsmodellen vor allem stochastische Zell-Zell-Variabilität berücksichtigen, wohingegen determinitische Ursachen, wie beispielsweise die Anzahl von Teilungszyklen, die eine Zelle durchläuft, häufig nicht betrachtet werden. Des Weiteren fehlt es an rigorosen Parameterschätz- und Analyseansätzen. Um diese Lücken zu schließen werden in dieser Arbeit zwei verallgemeinerte Klassen von Populationsmodellen für die Beschreibung (a) der Signalverarbeitung in heterogenen Zellpopulationen und (b) von Proliferationsprozessen hergeleitet. Die Beschreibung dieser Modellklassen wird ergänzt durch die Entwicklung von effizienten Algorithmen für die Modellanalyse.

Einführung in das Thema

Einzelzell- und Zellpopulationsdaten

Am Anfang der meisten biologischen Forschungsprojekte stehen die Durchführung von Experimenten und die Erhebung von Messdaten. Dies kann mit einer Vielzahl verschiedener Methoden erfolgen, z.B. Mikroskopie, Durchflusszytometrie und Western Blot Analysen. Die Wahl der Methode ist abhängig von der biologischen Fragestellung sowie dem durchzuführenden Experiment, und entscheidet darüber, welche Informationen zur Verfügung stehen.

Mikroskopische Methoden sind prinzipiell in der Lage, die Konzentrationen von chemischen Substanzen zeitaufgelöst zu ermitteln. Demgegenüber steht die Durchflusszytometrie, die lediglich Momentaufnahmen einzelner Zellen bereitstellt, da diese Methode es nicht erlaubt Einzelzellen über die Zeit zu verfolgen. Zu guter Letzt ermöglicht die Western Blot Analyse die Bestimmung der zeitveränderlichen Durchschnittskonzentration innerhalb einer Zellpopulation, ohne Zellen einzeln aufzulösen.

Hinsichtlich der Studie von einzelnen Zellen sind mikroskopische Methoden, und die Art von Messdaten, die sie zur Verfügung stellen, klar zu bevorzugen. Unglücklicherweise ist jedoch die Anzahl der analysierten Zellen oft gering. Durchflußzytometer können leicht zehntausende von Zellen erfassen. Daher liefert sie wesentlich mehr Informationen hinsichtlich der Populationsstatistik. Nichtsdestotrotz wurden Durchflußzytometriedaten bisher kaum verwendet, um die Parameter von heterogenen Zellpopulationen zu ermitteln.

Mathematische Modellierung biologischer Systems

Mathematische Modelle sind weit verbreitete Werkzeuge in der Systembiologie. Sie ermöglichen eine quantitative Beschreibung von biologischen Prozessen, die Ermittlung der Relevanz von Signalwegskomponenten und den Vergleich von konkurrierenden Hypothesen. Die mathematischen Modelle können in zwei Gruppen unterteilt werden: Einzelzell- und Zellpopulationsmodelle.

Einzelzellmodelle beschreiben die Vorgänge in einzelnen Zellen auf Basis von biochemischen Reaktionen. Abhängig von der Molekülzahl können diese Reaktionen als stochastisch oder deterministisch modelliert werden. Daraus ergeben sich verschiedene Modellklassen, unter anderem zeitkontinuierliche Markov Prozesse sowie stochastische und gewöhnliche Differentialgleichungen. Insbesondere Letztere sind in der medizinischen und pharmakologischen Anwendung weit verbreitet. Sie berücksichtigen häufig mehrere hundert biochemische Komponenten und Reaktionen, und ermöglichen ein ganzheitliches Verständnis des untersuchten Prozesses.

Da Einzelzellmodelle allein nicht die Untersuchung von Populationsheterogenität erlauben, ist die Modellierung von Einzelzellen durch die Modellierung von Zellpopulationen komplementiert. Für die meisten stochastischen Einzellzellmodelle gibt es eine dazugehörige Beschreibung der Populationsdynamik, z.B. die chemische Mastergleichung und die Fokker-Planck Gleichung. Diese Modelle beschreiben die Wahrscheinlichkeit für bestimmte Zustände der Einzelzelle unter Berücksichtigung des Prozessrauschens. Abgesehen vom Prozessrauschen kann es deterministische Unterschiede zwischen Zellen geben, so ist z.B. die Proteinsynthese abhängig vom Zellzyklusstatus. Unglücklicherweise gibt es kaum Modelle, die diese deterministische Zell-Zell-Variabilität berücksichtigen.

Über die Signaltransduktion hinaus ist auch die Zellproliferation ein interessanter Prozess, um das Langzeitverhalten von Zellsystemen zu beschreiben. Hierfür wurden insbesondere strukturierte Populationsmodelle eingeführt, die die Größe der Gesamtpopulation sowie die Verteilung des Zellalters, des Zellvolumens oder der Konzentration verschiedener biochemischer Komponenten beschreiben. Diese Modelle sind oft in Form von partiellen Differentialgleichungen beschrieben und erlauben nicht die Berechnung einer analytischen Lösung. Um die zugehörige Lösung zu berechnen, sind aufwendige numerische Verfahren notwendig.

Parameterschätzung und Unsicherheitsanalyse

Obwohl Einzelzell- und Zellpopulationsmodelle hochgradig verschieden sind, teilen sie doch eine entscheidende Eigenschaft: die Notwendigkeit verlässlicher Parameterwerte. Unglücklicherweise können diese Parameter auf Grund von experimentellen Beschränkungen meist nicht direkt gemessen werden, sondern müssen aus den gemessenen Konzentrationen ermittelt werden.

Die Allgegenwart von Parameterschätzproblemen führt zur Entwicklung einer Vielzahl von Schätzverfahren und Optimierungsalgorithmen. Die Grundidee ist jeweils, die unbekannten Parameter so anzupassen, dass die Diskrepanz von Modellvorhersage und Messdaten minimal wird. Obwohl dies einfach erscheint, erweist es sich jedoch häufig als anspruchsvoll. Abhängig von der Datenlage und der Modellstruktur kann das Optimierungsproblem lokale Minima besitzen, was das Finden des besten Parameters erschwert, und die Parameter können nicht identifizierbar sein. Daher ist eine Prüfung der Modellunsicherheit und der Vorhersagekraft essentiell.

Für Einzelzellmodelle wurden verschiedene Methoden entwickelt, die sowohl eine Parameterschätzung als auch eine Unsicherheitsanalyse ermöglichen. Solche Methoden fehlen für Populationsmodelle leider bisher. Eine Ursache ist die Vielzahl von möglichen Kombinationen von Modellen und Datentypen, eine andere ist der hohe Rechenaufwand, der mit der Simulation von Populationsmodellen verbunden ist. Insbesondere für die Integration der hochinformativen Zytometriedaten in Populationsmodelle gibt es bisher kaum Ansätze.

Forschungsbeiträge und Gliederung der Arbeit

In dieser Arbeit werden verallgemeinerte Modellierungsansätze und Bayes'sche Parameterschätzmethoden für (a) die Signalverarbeitung in und (b) die Proliferation von heterogenen Zellpopulation entwickelt. Die vorgestellten Methoden ermöglichen die Bestimmung von Modellparametern und deren Unsicherheiten unter Verwendung von Durchflusszytometriedaten. Des Weiteren ist die Bewertung der Vorhersagekraft der Modelle möglich.

Da die Simulation von Populationsmodellen potentiell rechenaufwendig ist, achten wir bei der Formulierung des Parameterschätzproblems auf deren Vermeidung. Dazu werden die beiden Modelle im Detail analysiert. Die jeweils gewonnen Erkenntnisse werden dazu genutzt, die Simulation und die Auswertung der Likelihood-Funktion zu verbinden.

Der Inhalt der Arbeit ist wie folgt gegliedert:

Kapitel 1 – Einleitung In diesem Kapitel führen wir die drei gebräuchlichsten Datentypen ein und diskutieren ihren Informationsgehalt, ihre statistischen Eigenschaften sowie experimentelle Einschränkungen. Außerdem beleuchten wir die Entstehungsgeschichte von

Einzelzell- und Populationsmodellen, und diskutieren ihre Anwendungen. Nachfolgend umreißen wir die verbreitetsten Parameterschätzmethoden und erörtern, warum die Anpassung von Einzelzellmodellen an Populationsdaten irreführende Ergebnisse liefern kann.

Kapitel 2 – Grundlagen Aufbauend auf Kapitel 1 werden hier biochemische Reaktionsnetzwerke formal eingeführt. Darauf basierend werden verbreitete Modellierungsansätze für Einzelzellen, die zeit-kontinuierlichen Markov Prozesse, die „chemical Langevin equation" und die „reaction rate equation", präsentiert. Für die stochastischen Modellierungsansätze darunter wird das statistische Modell vorgestellt. Dies schlägt die Brücke zur Populationsmodellierung, im Rahmen derer auch existierende Modelle für Zellproliferation umrissen werden. Abschließend behandeln wir Methoden für die Bayes'sche Parameterschätzung, unter anderem „Markov chain Monte Carlo" Methoden.

Kapitel 3 – Signalübertragung in heterogenen Zellpopulationen In Kapitel 3 entwickeln wir ein verallgemeinertes Modell für Signaltransduktionsprozesse in heterogenen Zellpopulationen. Dieses Modell berücksichtigt die Stochastizität der biochemischen Reaktionen sowie deterministische Unterschiede zwischen einzelnen Zellen, hier modelliert in Form von Parameterunterschieden. Der Signalübertragungsprozess kann für die Zellpopulation als erweiterte Fokker-Planck Gleichung formuliert werden. Um die näherungsweise Lösung dieser hochdimensionalen partiellen Differentialgleichungen zu ermitteln, kombinieren wir „direct sampling" mit Kerndichteschätzung.

Der zweite Teil des Kapitels beschäftigt sich mit der Schätzung der Parameterverteilung, die den deterministischen Anteil der Populationsheterogenität beschreibt. Eine Likelihood Funktion wird hergeleitet und, unter Verwendung eines affin parametrisierten Modells und des Superpositionsprinzips, in einen analytischen Ausdruck umformuliert. Die einzelnen Terme der sich ergebenden parametrischen Likelihood Funktion können vor der eigentlichen Optimierung bzw. der Unsicherheitsanalyse bestimmt werden. Das steigert die Effizienz. Außerdem beweisen wir, dass das sich ergebende Optimierungsproblem konvex und die A-posteriori-Verteilung unimodal ist. Dies erlaubt die Verwendung von effizienten Berechnungsverfahren, was wir anhand eines Populationsmodells für die TNF-induzierte apoptotische Signalübertragung illustrieren.

Kapitel 4 – Proliferation von heterogenen Zellpopulationen Um die Analyse von Signaltransduktionsprozessen zu komplementieren, wird in Kapitel 4 die Proliferation von heterogenen Zellpopulationen untersucht. Hierfür stellen wir ein strukturiertes Populationsmodell vor, das die Anzahl von Teilungen, die eine Zelle durchlaufen hat, sowie die Konzentration eines Markers berücksichtigt. Für dieses Modell, das bereits existierende Modelle verallgemeinert, stellen wir einen Zerlegungsansatz vor, der die Überführung des gekoppelten Systems von partiellen Differentialgleichungen in ein gekoppeltes System von gewöhnlichen Differentialgleichungen und eine Menge von analytisch lösbaren partiellen Differentialgleichungen ermöglicht. Ferner zeigen wir, dass bereits eine Berücksichtigung recht weniger Zustände der gewöhnlichen Differentialgleichungen eine gute Näherung der Gesamtlösung erlaubt.

Nachfolgend betrachten wir die zugehörige Parameterschätzung unter Verwendung von gebinnten Durchflusszytometriedaten. Hierfür ermitteln wir die Likelihood Funktion und zeigen eine Methode auf, diese auszuwerten, ohne das Modell vollständig zu simulieren. Dies reduziert den Rechenaufwand signifikant und erlaubt die Entwicklung globaler Optimierungs-

und Analysemethoden für das betrachtete Proliferationsmodell. Das Potential dieses Ansatzes wird anhand eines Datensatzes für T Lymphozyten illustriert.

Kapitel 5 – Fazit Die Resultate der Arbeit werden in Kapitel 5 zusammengefasst und in einen größeren Kontext gesetzt. Darüber hinaus diskutieren wir potentielle Anwendungen der entwickelten Methoden, offene Probleme und Ideen.

1 Introduction

1.1 Research motivation

Most biological research aims at achieving a mechanistic understanding of intracellular processes. To facilitate this, systems biology employs mathematical models. These models enable the integration of data from different data sources and experiments, rendering a holistic understanding of the process feasible. In particular, if model hypotheses have to be compared, mathematical models have been proven to be valuable.

The development of quantitative models relies on three key ingredients: the derivation of an appropriate mathematical model, the selection of the model parameters, and the model analysis. All three ingredients are essential, however, the latter two strongly depend on the derived model – especially the selected modeling framework. Depending on the biological question of interest, a multitude of different modeling frameworks may be employed.

In systems biology, most frequently single cell models are used, which describe a "typical" cell within a population. For these models a variety of parameter estimation methods and analysis tools exist. Unfortunately, experimental data used to estimate the parameters of these models are in general obtained using experiments which average over the population, e.g., western blotting. If the considered population is highly heterogeneous, e.g., due to stochastic gene expression, cell cycle state, or epigenetic differences, estimating the parameters of a single cell model from cell population data will most likely yield biologically misleading results. This is because the single cell model does not provide a mechanistic description for the cell population.

To understand the behavior of heterogeneous cell populations, it is crucial to develop cell population models. These models have to capture the dynamics of the overall population on the basis of the dynamics of the individual cells. In addition, to ensure predictive power of these models, parameter estimation methods and analysis methods are necessary – similar to the methods for single cell models.

In the following, we will argue that the available classes of population models mainly cover stochastic cell-to-cell variability, while deterministic sources, such as the number of cell divisions a cell underwent, are often disregarded. In addition, rigorous estimation and analysis approaches for complex population models are missing. To fill these gaps, this thesis provides two generalized modeling frameworks for (a) signal transduction in cell populations exhibiting parametric heterogeneity and (b) proliferating cell populations. The derivation of these model classes will be complemented by efficient, problem-specific parameter estimation and uncertainty analysis frameworks employing population level information. The focus is on intelligent methods exploiting the model structure and allowing for global exploration of the parameter space.

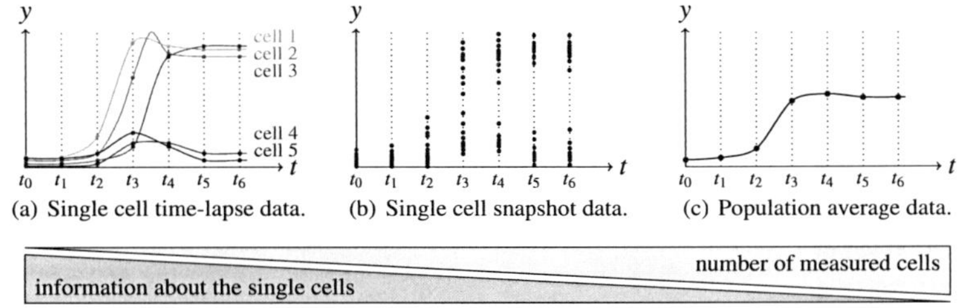

(a) Single cell time-lapse data. (b) Single cell snapshot data. (c) Population average data.

(d) Information content of the data types.

Figure 1.1: Illustration of the three most common types of measurement data: (a) single cell time-lapse data; (b) single cell snapshot data; and (c) population average. (d) Overview about the information content of the data types.

1.2 Research topic overview

1.2.1 Single cell and cell population data

The starting point of most biological research – also systems biological research – are experiments and the gathered measurement data. Depending on the biological system as well as the biological question, a multitude of different measurement techniques may be applied. Among others, these measurement techniques differ with respect to the measured quantity (e.g., mRNA or protein concentration) and the expected measurement error. Nevertheless, the key characteristics of the data are often highly similar allowing for a classification in three data types: single cell time-lapse data, single cell snapshot data, and population average data (see Figure 1.1)

Single cell time-lapse data provide information about the time-dependent properties (dynamics) of individual cells, as depicted in Figure 1.1(a). These data are typically obtained using transfected cells expressing a fluorescent protein, e.g., the green fluorescent protein (Tsien, 1998). Using fluorescently labeled cell lines, fluorescence time-lapse microscopy is performed, enabling the tracking of the concentration as well as the localization of the protein (Pepperkok & Ellenberg, 2006; Schroeder, 2011). Apparently, single cell time-lapse data are highly informative and allow for an in-depth analysis of the single cell dynamics as well as the population dynamics (see, e.g., (Tay *et al.*, 2010)). Unfortunately, the difficulty of establishing transfected cell lines and the potentially altered signaling caused be the transfection limits the availability of single cell time-lapse data. Furthermore, the often small number of measured single cell trajectories frequently provide an insufficient statistic of the population. Both problems are less pronounced for most measurement devices generating single cell snapshot data.

Single cell snapshot data provide information about the dynamics of the cell population with single cell resolution (see Figure 1.1(b)). The main difference to single cell time-lapse data is that individual cells are not tracked over time but merely measured at one time instance. Therefore, single cell snapshot data are less informative than single cell time-lapse data with respect to the single cell dynamics. On the other hand, the statistics are better as the number of cells measured at each time instance is typically larger. The most common devices providing single cell snapshot data are flow cytometers (Herzenberg *et al.*, 2006), which can be used to study transfected and stained cells (Hawkins *et al.*, 2007; Lyons & Parish, 1994). In typical measurements tens of thousands of cells are analyzed, providing a rich data source which may still allow for classification of subpopulations, like single cell time-lapse data (Andres *et al.*, 2012).

Note that single cell snapshot data are often stored as histograms, which prohibit a retrieval of the precise single cell data. Still, if the histogram bins (in biology also often called channels) are narrow, histograms are almost as informative as the original single-cell data. If the histogram bins are wide, an information loss occurs like for the population average data.

Population average data provide information about the averaged dynamics of the overall population (see Figure 1.1(c)). Unlike the previous data types, population average data do not have single cell resolution. Therefore, the structure of the population is hidden. It is, for instance, not possible to distinguish between a strong response of a small subpopulation and a weak response of the whole population, as the average response in the population might be identical for those two scenarios. Despite this drawback, population average data are more common than the other data types. This might be due to the easier experimental setups. Population average data can be collected with a variety of devices, among others by Western blots (Renart *et al.*, 1979) and microarrays (Malone & Oliver, 2011).

As the different data types carry different information (see Figure 1.1(d)), different analysis tools have been developed. These tools range from pure data analysis via histograms (Lampariello, 2009; Lampariello & Aiello, 1998; Watson, 2001), kernel density estimators (Andres *et al.*, 2012), and Gaussian mixture models (Hyrien & Zand, 2008; Song *et al.*, 2010; Wang & Huang, 2007), to more model-based approaches. The model-based approaches have in this case the advantages that data originating from different experiments and data sources can be considered simultaneously.

1.2.2 Mathematical modeling of biological systems

Mathematical tools have supported biological research for decades, starting with the statistical analysis of data (Fay & Proschan, 2010; Good, 2004; Huang, 2010; Welch, 1947). After the groundbreaking works by Hodgkin & Huxley (1952), Turing (1952), and von Foerster (1959), also the application of mathematical models has become increasingly common. Nowadays, mechanistic models are widespread in mathematical, theoretical, as well as systems biology. They enable the quantitive description of biological systems, the assessment of key-players in biochemical networks, and the comparison of competing model hypotheses.

The mathematical models used in cell biology can be roughly separated in two classes: single cell models, and population models. These classes are strongly related (as we discuss in Chapter 2) and split up in several subclasses, such as stochastic (Higham, 2008; Wilkinson, 2009) or deterministic models (Klipp *et al.*, 2005a).

Mathematical modeling of single cells

Single cell models describe the behavior of individual cells. They can be mechanistic or phenomenological in nature. Mechanistic single cell models are based upon chemical reaction kinetics, such as the law of mass action (Klipp *et al.*, 2005a) or enzyme kinetics (Michaelis & Menten, 1913). Phenomenological single cell models employ qualitative descriptions of the interactions of different biological entities (Angeli *et al.*, 2004; Chaves *et al.*, 2008; Schlatter *et al.*, 2009). Mechanistic as well as phenomenological single cell models might consider the stochasticity of biochemical reactions due to small molecule numbers (Gillespie, 1992), and noisy processes, such as gene expression (Elowitz *et al.*, 2002).

The most commonly used mathematical modeling frameworks are Markov jump processes, chemical Langevin equations, and deterministic chemical kinetic models. These models enable the description of a variety of processes, i.e., metabolism, gene regulation, and signal transduction. In recent years, increased amounts of quantitative data enabled the development of high-dimensional single cell models accounting for hundreds of biochemical species and reactions (Albeck *et al.*, 2008a; Eissing *et al.*, 2011; Klipp *et al.*, 2005b; Schöberl *et al.*, 2002, 2009). These models allow for the holistic understanding of complex biological processes in a highly quantitative manner and therefore find many application in, e.g., pharmacology or cancer research.

Mathematical modeling of cell populations

The research on single cell modeling is complemented by modeling cell populations. For most stochastic single cell models of signal transduction a corresponding mathematical description of the populations dynamics exist (Gillespie, 1992, 2000), e.g., the Chemical Master Equation and the Fokker-Planck equation. These population models describe the statistics of the individual cells, i.e., the probability (density) of occupying a particular state. Besides the intrinsic stochasticity, also extrinsic sources of cell-to-cell variability (Huh & Paulsson, 2011; Swain *et al.*, 2002) and regulated cell-to-cell variability (Snijder & Pelkmans, 2011) are biologically important. Unfortunately, hardly any quantitative models are available which account for those. Merely some first attempts have been made in this respect (see, e.g., (Spencer *et al.*, 2009)).

Beyond signaling also cell proliferation and cell death are of interest. Therefore, population models have been introduced including these phenomena and governing the long-term behavior of populations. One of the first population models describing a simple proliferation process has been developed by von Foerster (1959). This well-known work has been generalized, for instance, in (Diekmann *et al.*, 1998; Gyllenberg, 1986; Tsuchiya *et al.*, 1966), to account for the structure within the population. These generalizations are biologically necessary, but the corresponding models suffer a tremendously increased complexity.

While for the von Foerster model (Sinko & Streifer, 1967; Trucco, 1965) and a few other population models (Jahnke & Huisinga, 2007; Weiße *et al.*, 2010) a closed form solution has been found, the quantitative dynamics of most population models can only be assessed via numerical simulation. As the models are either high-dimensional systems of ODEs or PDEs, this is problematic and requires sophisticated numerical schemes which renders an in-depth analysis difficult.

An in-depth discussion of single cell models, population models and their relation is provided in Chapter 2. Though single cell models and population models are greatly different, they share one key characteristic: the need for accurate parameters values.

1.2.3 Parameter estimation and uncertainty analysis

Most model parameters cannot be measured directly due to the limitations of experimental techniques and devices. These unknown parameters have to be inferred from measurement data using parameter estimation (Ljung, 1999), also called parameter inference.

In recent years, parameter estimation has gained much attention due to its ubiquity (Engl *et al.*, 2009). This has manifested in the development of a wide array of parameter estimation procedures and optimization algorithms. The key idea behind parameter estimation is to adapt the unknown parameters such that the model predictions match the experimentally observed data. While this sounds simple, parameter estimation is often challenging. Depending on the available measurement data, the inverse problem of inferring the parameters from the data may be ill-posed (Engl *et al.*, 2009). This means that because of the structural properties of systems (Audoly *et al.*, 2001; Cobelli & DiStefano, 1980; Raue *et al.*, 2009) and/or limitation of the data (Apgar *et al.*, 2010; Gutenkunst *et al.*, 2007; Hengl *et al.*, 2007; Komorowski *et al.*, 2011), e.g., caused by low time resolution and measurement noise, some parameters may be non-identifiable. Thus, in addition to sophisticated parameter estimation procedures, a rigorous assessment of the parameter and prediction uncertainty remaining after the integration of the data into the model – often called identifiability or uncertainty analysis – is important.

Parameter estimation and uncertainty analysis for single-cell models

Most methods for parameter estimation and uncertainty analysis have been developed for deterministic and stochastic single-cell models. In particular, maximum likelihood formulations (Brännmark *et al.*, 2010; Brown *et al.*, 2004; Chen *et al.*, 2010; Huys & Paninski, 2009; Maiwald & Timmer, 2008; Reinker *et al.*, 2006; Weber *et al.*, 2011) and Bayesian formulations (Busetto & Buhmann, 2009; Busetto *et al.*, 2009; Dargatz, 2010; Golightly & Wilkinson, 2010; Klinke, 2009; Toni *et al.*, 2009; Wilkinson, 2007, 2010) are widely used. Less frequently also set-based estimation (Kieffer & Walter, 2011; Küpfer *et al.*, 2007; Rumschinski *et al.*, 2010; Hasenauer *et al.*, 2010c), moment matching (Lillacci & Khammash, 2010a,b), and nonlinear observers (Fey & Bullinger, 2009; Quach *et al.*, 2007) are applied. As in many cases, the likelihood function, which is a measure for the distance between model prediction and data, can be evaluated efficiently, a global search in the parameter space for the optimal parameter is feasible.

Depending on the parameter estimation approach, distinct methods for identifiability and uncertainty analysis are employed. Sample-based credibility intervals are the tool of choice when using a Bayesian framework, as they do not require additional computations. For maximum likelihood estimation confidence intervals computed via local approximations (Balsa-Canto *et al.*, 2010; Joshi *et al.*, 2006), bootstrapping (Carpenter & Bithell, 2000; Joshi *et al.*, 2006), and profile likelihoods (Meeker & Escobar, 1995; Raue *et al.*, 2009, 2011; Weber *et al.*, 2011) are most common.

The variety of the used methods shows that there is no gold standard for parameter estimation and uncertainty analysis of single cell models. Still, there are methods (Balsa-Canto *et al.*, 2008; Moles *et al.*, 2003) and even toolboxes (Hoops *et al.*, 2006; Maiwald & Timmer, 2008; Schmidt & Jirstrand, 2006) which have proven to be reliable.

The main problem for single cell parameter estimation is that most experimental data are obtained using population average experiments, such as western blotting. If the considered population is heterogeneous, fitting a single cell model to cell population data yields biologically misleading results. In particular, if decision processes, such as apoptosis or differentiation, are considered where some cells respond and others maintain their state (as illustrated in Figure 1.1(a)), single cell models fitted to the population average describe none of the subpopulations and a biological sound interpretation of the estimation result is hardly possible. If parameter estimation for a single cell is performed, it has to be ensured that either solely single cell time-lapse data are used or that the population is homogeneous. As single cell time-lapses are rarely available and as even clonal populations are heterogeneous with respect to, e.g., cell cycle state, cell age and protein abundance, population models become more common and are the next logical step.

Parameter estimation and uncertainty analysis for population models

Parameter estimation and uncertainty analysis for population models has many facets as highly problem-specific formulations of the problem are employed for distinct combinations of population models and data types. This involves primarily differences of the likelihood function, which will be outlined in the following for the different data types.

As single cell time-lapse data enable the assessment of the complete single cell dynamics, frequently single cell estimation schemes are applied (Kalita *et al.*, 2011). These schemes rely on the assumption that each cell possesses its own realization of the parameter value. Recently, this approach has been generalized to enable the parameterization of the overall probability distribution (Koeppl *et al.*, 2012). This results in an indirect coupling of the estimation problems for the individual cells. While such hierarchical approaches – borrowed from mixed effect modeling in pharmacokinetics (Al-Banna *et al.*, 1990) – are powerful, merely the simultaneous analysis of few cells is feasible as the dimensionality of the estimation problem increases linearly with the number of cells. This limits the statistical reliability of the models. Furthermore, the information about the individual cells must be rich to avoid severe identifiability problems. For this reason, hierarchical models and models with single cell resolution are not used to study single cell snapshot data and population average data containing thousands of measured cells.

Single cell snapshot data are mainly used to train population models for signal transduction or proliferation processes. While the data type is identical, the models have distinct characteristics resulting in different formulation of the parameter estimation problems. For signaling pathway models, maximum likelihood and Bayesian formulations of the estimation problem are known (Nüesch, 2010), however, most commonly norm-based approaches are used due to their simplicity (Munsky & Khammash, 2010; Munsky *et al.*, 2009), although they are statistically not well motivated. For proliferation models on the other hand, only heuristic norm-based formulations based on least-squares and generalized least-squares (Banks *et al.*, 2011, 2010; Luzyanina *et al.*, 2007b) have been published. While these norm-based approaches neglect the underlying statistics of the measurement process, also none of the likelihood-based estimation procedures takes the measurement error on the single cell level or the limited resolution of the measurement device into account. Furthermore, for both model classes no estimation procedures are available which consider deterministic cell-to-cell variability.

While this already shows that parameter estimation methods employing single cell time-lapse data and single cell snapshot data are not well developed, the situation is even worse for population average data. To our knowledge, population average data have not been used to infer information about population models but merely to estimate the parameters of single-cell models. The reason might be the strongly decreased information content of average data compared to data with single cell resolution.

On the methodological side it has to be mentioned that for most of the above estimation methods, no methods for uncertainty analysis are available. Merely the few Bayesian methods (Kalita *et al.*, 2011; Koeppl *et al.*, 2012; Nüesch, 2010) provide statistics about the parameter uncertainty. Others do not even allow a global exploration of the parameter space but employ local optimization algorithms.

1.2.4 Methodological challenges

In the previous sections an overview about state-of-the-art approaches for modeling, parameter estimation, and uncertainty analysis has been provided. While it might be claimed that methods for single cells are well developed, we apparently lack sophisticated methods for cell population. This is due to three key challenges: First, models of cell populations have to account for different sources of cell-to-cell variability (of stochastic and deterministic nature). Second, consistent likelihood functions for population experiments are complex. Third, simulation of population models – and therefore also the evaluation of the likelihood functions – is time consuming. These are the main reasons for the lack of reliable methods for the population model as they prevent the transfer of methods develop for single cell models to cell population models.

1.3 Contribution of this thesis

The challenges of modeling, parameter estimation and uncertainty analysis of cell populations are addressed in this thesis for two classes of processes: signaling in and proliferation of heterogeneous cell populations. For both processes, model development and parameter estimation – two mostly separate steps – are intertwined. We show that this intertwining increases the computational efficiency while simultaneously enabling the study of more complex models. The resulting population models are more consistent with the biological process and, from various points of view, generalizations of existing models. Throughout the thesis, we employ different types of single cell snapshot data for parameter estimation, thereby providing a first rigorous statistical treatment of these highly informative experimental setups.

In the first part of this thesis, we present a

- mechanistic modeling framework for signal processing in heterogeneous cell populations, accounting for intrinsic and extrinsic cell-to-cell variability.

Intrinsic variability is model by stochasticity of biochemical reactions, while extrinsic (or regulated) variability is modeled by differences in parameter values and initial conditions among individual cells. We show that the dynamics of such heterogeneous cell populations are governed by an augmented Fokker-Planck equation. For this model class we propose a

- parameter estimation method, which allows for the reconstruction of the extrinsic cell-to-cell variability from single cell snapshot data.

This method employs a simplified representation of the likelihood function, which is derived using a linear parameterization of the parameter distribution and the fact that the augmented Fokker-Planck equation fulfills the superposition principle. Using both, the complex likelihood function can be reformulated to an algebraic expression, which can be evaluated efficiently, enabling the global parameter search and uncertainty analysis. We even verify that the optimum is the solution of a quasiconvex optimization problem.

In the second part of this thesis, we propose a

- generalized structured population model for cell proliferation, which incorporates discrete age structure and continuous label dynamics.

We prove that under mild assumptions, the corresponding system of coupled partial differential equations can be decomposed into a system of ordinary differential equations and a set of decoupled PDEs. This reduces the computational effort drastically. Employing this generalized model and an efficient decomposition-based solution scheme, we approach the estimation of the unknown model parameters. For this purpose, we derive a

- consistent likelihood function for binned single-cell snapshot data,

by formulating the measurement process as generalized Bernoulli trial. Employing the structure of the model, we show that the

- evaluation of the likelihood function of the coupled PDE model merely requires the simulation of a low-dimensional ODE.

This results in a speed-up of several orders of magnitude compared to existing formulations.

In both parts of this thesis, Bayesian formulations are employed for parameter estimation, as well as uncertainty analysis. While Bayesian methods are not commonly used in population biology (and with PDE models in general), we prove that appropriate (re-)formulations and modifications of model and likelihood functions renders their application computationally trackable. These modifications employ common control engineering tools and respect the particular context of the model, especially showing that states and outputs should be distinguished when studying population models.

Employing the methods developed in this thesis, a rigorous modeling and a consistent parameter estimation and uncertainty analysis for population models becomes feasible. This allows for the study of quantities which cannot be measured.

1.4 Outline of this thesis

To set the ground for this thesis, **Chapter 2** provides an introduction to stochastic and deterministic modeling in biology. The prevalent modeling approaches for single cells and cell populations are presented and their interrelations are discussed. In addition, this chapter contains an introduction to Bayesian parameter estimations and statistical analysis tools, such as Markov chain Monte Carlo sampling and Bayesian confidence intervals.

In **Chapter 3**, a generalized model for signal transduction in heterogeneous cell population is introduced and analyzed. Furthermore, a numerical solution algorithm is proposed. Secondly, the likelihood function is derived and simplified by employing the density parameterization. For the simplified likelihood function the parameter estimation and uncertainty analysis problems are formulated. All methods are applied to a model for TNF-induced apoptotic signaling.

The analysis of signaling processes is complemented by the study of proliferation processes. In **Chapter 4** a division- and label-structured population model is introduced. For this model a decomposition approach is provided to reduced the computational complexity. Next, the measurement process is modeled and the likelihood function derived. This likelihood function is intertwined with the decomposed model to enable efficient Bayesian parameter estimation. The whole approach is employed to study T lymphocytes proliferation.

Chapter 5 provides a summary of the key results and discusses them in the context of existing methods. The thesis is concluded by outlining potential applications of the introduced concepts, open problems, and ideas.

2 Background

In the last chapter, the topic and the contribution of this thesis has been outlined. To provide the background knowledge necessary in the following, this chapter gives a survey of quantitative mathematical modeling in biology as well as Bayesian parameter estimation. In Section 2.1, the notion of biochemical reaction networks is introduced along with the three most common modeling approaches in quantitative systems biology: Markov jump processes, chemical Langevin equation and reaction rate equation. The section on single cell modeling is complemented by an overview about modeling of cell populations in Section 2.2. We outline ensemble and density-based population modeling. Finally, in Section 2.3, we discuss the fundamentals of Bayesian parameter estimation, including computational methods, such as Markov chain Monte Carlo sampling.

2.1 Single cells models

Modeling of single cell dynamics has been a field of active research for several decades, starting with the groundbreaking work of Hodgkin & Huxley (1952), who described the dynamics of individual neurons using ordinary differential equations (ODEs). In the course of time, their work has been extended to other fields, i.e., metabolism, signal transduction, and gene regulation (Klipp *et al.*, 2005a). Nowadays, there exists a variety of different modeling approaches besides ODEs, which take, e.g., stochastic or spatial effects into account. All of these modeling approaches share two essential elements, chemical species $(S_1, S_2, \ldots, S_{d_s})$ and chemical reactions $(R_1, R_2, \ldots, R_{d_r})$.

A chemical species is an ensemble of chemically identical molecular entities, such as proteins and RNA molecules, while a process, which results in the interconversion of chemical species, is referred to as chemical reaction (McNaught & Wilkinson, 1997), e.g., synthesis, degradation, and phosphorylation. Accordingly, chemical reactions are defined via lists of reactants (r) and products (p), and can be written as:

$$R_j: \quad \sum_{i=1}^{d_s} \nu_{ij}^{(r)} S_i \rightarrow \sum_{i=1}^{d_s} \nu_{ij}^{(p)} S_i, \quad j = 1, \ldots, d_r. \tag{2.1}$$

Thereby, $\nu_{ij}^{(r)}$ $(\nu_{ij}^{(p)}) \in \mathbb{N}_0$ is the stoichiometric coefficient of species i in reaction j, which denotes the number of molecules consumed (produced) when the reaction takes place (Klipp *et al.*, 2005a). The overall stoichiometry of reaction j is

$$\nu_j = \left(\nu_{ij}^{(p)} - \nu_{ij}^{(r)}\right)_{i=1,\ldots,d_s}, \quad j = 1, \ldots, d_r. \tag{2.2}$$

in which $\nu_j \in \mathbb{Z}$ is the net interconversion of species.

Based upon the notion of chemical species and chemical reactions, we present the three most common modeling approaches for single cells: Markov jump processes, chemical Langevin equations, and reaction rate equations.

Table 2.1: Propensity functions for reaction following mass action kinetics (Gillespie, 1977).

Reaction type	Propensity function
$\emptyset \rightarrow$ product	$a_j(X, \theta) = \theta$
$S_i \rightarrow$ product	$a_j(X, \theta) = \theta X_i$
$S_{i_1} + S_{i_2} \rightarrow$ product	$a_j(X, \theta) = \theta X_{i_1} X_{i_2}$
$2S_i \rightarrow$ product	$a_j(X, \theta) = \frac{\theta}{2} X_i(X_i - 1)$

2.1.1 Stochastic chemical kinetic models

Markov jump process

Molecules are discrete entities and reactions are discrete events. For this reason, the most natural way to describe biochemical reaction networks are discrete-state continuous-time Markov processes (Gillespie, 1977; Higham, 2008), also called Markov jump processes (MJP).

The state, $X_t \in \mathbb{N}_0^{d_s}$, of a Markov jump process at time t is the collection of the ensemble sizes of the individual chemical species at time t, $X_{i,t} \in \mathbb{N}_0$,

$$X_t = (X_{i,t})_{i=1,\ldots,d_s}, \tag{2.3}$$

and thus an integer-valued random vector. The state of a MJP remains constant as long as no reaction takes place. If the j-th chemical reaction, R_j, occurs the ensemble sizes changes according to the stoichiometry of the reaction,

$$X_t^+ = X_t^- + \nu_j, \tag{2.4}$$

in which X_t^- and X_t^+ are the ensemble size directly before and directly after the reaction, respectively (Gillespie, 1977). The index j of the next reaction as well as the time to the next reaction are randomly distributed (Feller, 1940). Both depend directly on the propensity functions $a_j(X_t, \theta)$, which are associated to the speed of a reaction. The probability that reaction R_j fires in the next infinitesimal time interval dt is given by $a_j(X_t, \theta)dt$ (Gillespie, 1977), yielding

$$P(X_{t+dt} = x + \nu_j | X_t = x) = a_j(X_t, \theta)dt, \quad j = 1, \ldots, d_r. \tag{2.5}$$

The propensity $a_j(X_t, \theta)$ depends on the reaction rate constants, $\theta \in \mathbb{R}_+^{d_\theta}$, and the current state (Gillespie, 1977). Some examples of common propensity functions can be found in Table 2.1.

Markov jump processes can be viewed as random walks on a graph. The nodes of this graph represent different combinations of ensemble sizes, while directed edges are possible transitions, whose transition probabilities are associated to the propensity functions. This is illustrated in Figure 2.1 for the well-known birth-death process (Wilkinson, 2009) with reactions $R_1 : \emptyset \xrightarrow{k} X$ and $R_2 : X \xrightarrow{\gamma} \emptyset$, propensity functions $a_1(X, \theta) = k$ and $a_2(X, \theta) = \gamma X$, and parameter vector $\theta = [k, \gamma]^T$. As Markov models accounts explicitly for this discrete nature of events and molecule numbers, they reflect exactly the intrinsic noise and stochasticity of the biochemical processes.

For the simulation of MJPs, a variety of algorithms are available. The most commonly used one is Gillespie's algorithm (Gillespie, 1977, 1992), which allows for an exact stochastic

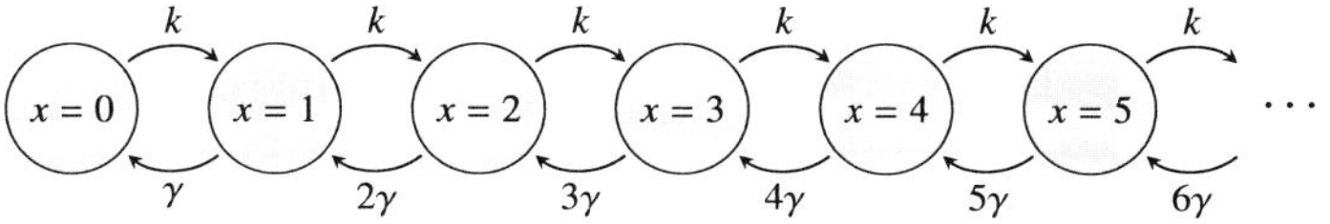

Figure 2.1: Illustration of discrete-state continuous-time Markov chains for the birth-death process. Each individual realization of the Markov chain corresponds to a random walk on this graph.

simulation. Unfortunately, if the number of reaction events is large, i.e., due to large molecule number and/or fast reaction kinetics, the computational cost of Gillespie's algorithm becomes intractable. To circumvent this, different approximate stochastic simulation algorithm have been introduced, employing, for instance, τ-leaping (Gillespie, 2001; Rathinam *et al.*, 2003) or systems-partitioning (Rao & Arkin, 2003). Still, there are many systems whose simulation is very time consuming even with sophisticated methods (Wilkinson, 2009).

Chemical Langevin equation

To overcome the computational effort of simulating the Markov jump process, the chemical Langevin equation (CLE) can be used. The CLE is a system of stochastic differential equations,

$$dX_t = \mu(X_t, \theta)dt + \sigma(X_t, \theta)dW_t, \quad X_0 = X(0), \tag{2.6}$$

in which $X_t \in \mathbb{R}_+^{d_s}$ denotes the state vector and $W_t \in \mathbb{R}^{d_w}$ denotes a Wiener process. The drift function $\mu : \mathbb{R}_+^{d_s} \times \mathbb{R}_+^{d_\theta} \to \mathbb{R}^{d_s}$ and the diffusion function $\sigma : \mathbb{R}_+^{d_s} \times \mathbb{R}_+^{d_\theta} \to \mathbb{R}^{d_s \times d_w}$ determine the deterministic and stochastic contributions to the dynamics, respectively.

Different methods exist to approximate MJPs with CLEs, see Gillespie (2000), Hasty *et al.* (2000), Hasty *et al.* (2001), Kepler & Elston (2001), Wilkinson (2009) and references therein. The most appealing approach is the diffusion approximation (Gillespie, 2000), establishing a link between the propensity function of the MJP and the deterministic and stochastic component of the vector field of the CLE. Applying the diffusion approximation of the MJP, the chemical Langevin equation becomes

$$dX_t = \sum_{j=1}^{d_r} \nu_j a_j(X_t, \theta)dt + \sum_{j=1}^{d_r} \nu_j \sqrt{a_j(X_t, \theta)}dW_{j,t}, \tag{2.7}$$

as proven in (Gillespie, 2000). Hence, the drift and the diffusion function are completely determined by the propensity functions and given by $\mu(X_t, \theta) = \sum_{j=1}^{d_r} \nu_j a(X_t, \theta)$ and $\sigma(X_t, \theta) = (\nu_1, \ldots, \nu_r) \sqrt{a(X_t, \theta)}$, with $d_w = d_r$. In contrast to ad hoc methods (see, e.g., Hasty *et al.* (2001, 2000), Glauche *et al.* (2010), and Schittler *et al.* (2010)), the diffusion approximation does not introduce additional parameters.

Still, the approximation of the MJP by a CLE introduces two disadvantages independent of the precise approximation scheme. Firstly, the statistical properties (e.g., first and second moment) of the MJP are not necessarily captured by the CLE, and secondly, the discrete nature of the MJP is lost as X_t is not longer integer-valued. In practice, the former might cause severe problems, for instance, when predicting the process dynamics. Fortunately, for many systems, the moments of the states are conserved by the CLE (Higham & Khanin, 2008; Kepler & Elston, 2001). The latter disadvantage of the approximation is also undesirable but allows for a computational speed-up and merely complicates the biological interpretation slightly.

The solution of the chemical Langevin equation can be computed using any stochastic differential equation solvers, such as the Euler-Marujama method (Higham, 2001; Kloeden & Platen, 1999), the Milstein method (Higham, 2001; Kloeden & Platen, 1999), or higher-order Runge-Kutta methods (Buckwar & Winkler, 2004, 2006). Therefore, the computational effort does not directly depend on the number of molecules or the number of reaction events, but merely on properties of the functions μ and σ, such as their stiffnesses. This results for many processes in a computational speed-up of several orders of magnitude. Still, systems with dozens of species challenge even sophisticated stochastic differential equation solvers, which is why deterministic models have to be used.

2.1.2 Reaction rate equation

The most common approach to chemical kinetic modeling is the reaction rate equation (RRE) (Fall *et al.*, 2010; Klipp *et al.*, 2005a), which is an ordinary differential equation. The RRE provides a description of the biochemical processes if the ensemble sizes of all chemical species per unit volume are sufficiently large. In this case, the protein abundance in a cell with volume Ω is a continuous quantity and the time-dependent species concentration is defined as

$$x(t) = \frac{X_t}{\Omega} \in \mathbb{R}_+^{d_s}. \tag{2.8}$$

If elements of X_t correspond to the molecule number in different cellular compartments, such as cytosol or nucleus, the scaling has to be performed with respect to the associated volume.

As the number of molecules is large, the concentration changes deterministically according to

$$\frac{dx}{dt} = \mu(x, \theta), \tag{2.9}$$

in which $\mu : \mathbb{R}_+^{d_s} \times \mathbb{R}_+^{d_\theta} \rightarrow \mathbb{R}^{d_s}$ denotes the vector field, respectively drift function, of the RRE. The drift function can be derived from the chemical reactions and their stoichiometry, yielding

$$\frac{dx}{dt} = \sum_{j=1}^{d_r} \nu_j \tilde{a}_j(x, \tilde{\theta}), \tag{2.10}$$

in which $\tilde{a}(x, \tilde{\theta}) \in \mathbb{R}_+^{d_r}$ and $\tilde{\theta} \in \mathbb{R}_+^{d_\theta}$ denote macroscopic reaction rates and macroscopic parameters, respectively. The macroscopic reaction rates $\tilde{a}_j(x, \tilde{\theta})$ and parameters $\tilde{\theta}$ are closely related to the propensity function $a_j(X_t, \theta)$ of the Markov jump process, via the limit of $\Omega\tilde{a}_j(x, \tilde{\theta}) = a_j(X_t, \theta)$ for $\Omega \rightarrow \infty$. An in-depth discussion of this and the relation of the parameters of deterministic and stochastic processes has been provided by Gillespie (2000). Common macroscopic rate laws are shown in Table 2.2.

As the RRE is purely deterministic, the intrinsic noise of the process is not captured. Furthermore, it cannot be guaranteed that the RRE describes the mean concentration of the underlying Markov jump process. While these two drawbacks of the RRE limits its application, the multitude of efficient ODE solvers and toolboxes (Maiwald & Timmer, 2008; Schmidt & Jirstrand, 2006) allows for the consideration of high-dimensional systems. Studies of

Table 2.2: Deterministic rate laws for reaction following mass action kinetics (Klipp *et al.*, 2005a).

Reaction type	Rate law
$\emptyset \to$ product	$\tilde{a}_j(x,\theta) = \theta$
$S_i \to$ product	$\tilde{a}_j(x,\theta) = \theta\, x_i$
$S_{i_1} + S_{i_2} \to$ product	$\tilde{a}_j(x,\theta) = \theta\, x_{i_1} x_{i_2}$
$2S_i \to$ product	$\tilde{a}_j(x,\theta) = \theta x_i^2$
$\sum_{i=1}^{d_s} \nu_{ij}^{(r)} S_i \to$ product	$\tilde{a}_j(x,\theta) = \theta \prod_{i=1}^{d_s} x_i^{\nu_{ij}^{(r)}}$

processes involving dozens or even hundreds of species become increasingly common, see, e.g., (Klipp *et al.*, 2005b; Schöberl *et al.*, 2002; Spencer *et al.*, 2009).

To conclude this overview about modeling approaches for single cells, we provide a simulation study of the birth-death process (Figure 2.1) in Figure 2.2. This study is similar to the one of Wilkinson (2009) and illustrates the properties of the different model classes, as well as the nature of the different processes: MJP – discrete state and stochastic dynamics; CLE – continuous state and stochastic dynamics; and RRE – continuous state and deterministic dynamics. While the MJP model is the best approximation to the underlying biochemical process, for this example, the RRE still describes the dynamics of the mean and can be evaluated at a low computational cost.

2.2 Cell population models

About at the same time as Hodgkin & Huxley (1952) introduced their first single cell model, the development of population models began. This beginning is marked by the well-known work of von Foerster (1959), who derived a partial differential equation model for the age-distribution in growing cell populations. Since then, the field of population modeling has been active and a variety of population models has been derived. Most of these models still focus on simple descriptions of cell division and study the resulting age-structure. Only during the last years, the usage of population models to study intracellular signaling and cell-to-cell variability became more prevalent.

In this section, we introduce the four most common classes of population models – cell ensemble models, population balance models, Chemical Master Equations, and Fokker-Planck equations – and their mathematical properties. We will start with ensemble-based approaches and end with density-based descriptions.

2.2.1 Cell ensemble models

Cell population consist of individual cells, which can be described using the stochastic and/or deterministic single cell models introduced above. Thus, the most intuitive cell populations model is a collection of a finite number of single cell models $\Sigma_{\text{cell}}^{(i)}$, each describing a member of the population. This results mathematically in the cell ensemble model,

$$\Sigma_{\text{pop}} = \left\{ \Sigma_{\text{cell}}^{i} | i = 1, \ldots, C \right\}, \tag{2.11}$$

Markov jump process

Model: $P(X_{t+dt} = x + 1 | X_t = x) = k dt$

$P(X_{t+dt} = x - 1 | X_t = x) = \gamma x dt$

Initial condition: $X_0 = 0$

Chemical Langevin equation

Model: $dX_t = (k - \gamma X_t)dt + \sqrt{k} dW_{1,t} - \sqrt{\gamma X_t} dW_{2,t}$

Initial condition: $X_0 = 0$

Reaction rate equation

Model: $\frac{dx}{dt} = k - \gamma x$

Initial condition: $x(0) = 0$

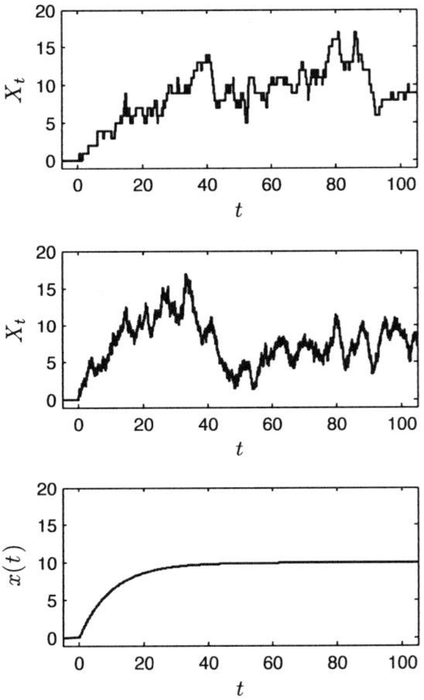

Figure 2.2: Modeling and simulation of the birth-death process with Markov jump model, chemical Langevin equation, and reaction rate equation. The mathematical description of the MJP, the CLE, and the RRE, as well as the corresponding simulation results are shown. For the simulations we choose the birth rate $k = 1$, death rate $\gamma = 0.1$, and the cell volume $\Omega = 1$ (similar to simulations in (Wilkinson, 2009)). The plots illustrate the discrete stochastic nature of the MJP, the continuous stochastic nature of the CLE, and the continuous deterministic nature of the RRE. While all three models describe the same process, the level of mechanistic detail strongly varies. The MJP resolves each reaction events, while the RRE merely describes the mean of the molecule number.

in which $C \in \mathbb{N}$ the size of the population. The single cell dynamics may be described by a chemical Langevin equation, a Markov jump process, or a reaction rate equation, and hence be stochastic as well as deterministic. Furthermore, cell-to-cell variability may be considered, for instance, by differences in parameters and initial conditions among individual cells (Albeck *et al.*, 2008b; Eissing *et al.*, 2004; Spencer *et al.*, 2009). Thus, the class of ensemble models is broad, allowing their application in fields, such as biochemical engineering (Henson *et al.*, 2002; Mantzaris, 2007), stem cell biology (Glauche *et al.*, 2009, 2011), immunology (Glauche *et al.*, 2009, 2011), as well as cancer research (Albeck *et al.*, 2008b; Brown *et al.*, 2010; Eissing *et al.*, 2004; Niepel *et al.*, 2009; Spencer *et al.*, 2009).

While the model introduced above assumes that the individual cells do not interact, also generalized frameworks are available. Such generalized ensemble models allow for communication via the extracellular medium, direct cell-cell interactions, and interaction with the environment (see (Anderson & Quaranta, 2008; Buske *et al.*, 2011; Lang *et al.*, 2011; Walker *et al.*, 2008) and references therein). To allow for these extensions the spatial position of the individual cell is taken into account as well as distribution processes.

Despite their generality, ensemble models are still underrepresented in quantitative systems biology. The reason for this is that the simulation of ensemble models is often computationally demanding, in particular if spatial effects are taken into account. Due to the high computational complexity it is often not possible to fit the models to available measurement data or to study the uncertainty of the model predictions. Even the use of local analysis methods, such as sensitivity analysis, might be rendered infeasible.

2.2.2 Population density models

To avoid the computational complexity of ensemble models, frequently density models are used. Density models provide a description of the population dynamics, without the need for the simulation of individual cells.

Population balance models

In the middle of the last century, experimental methods to study intracellular processes have been rather limited. Therefore, much of the biological research focused on biological problems, such as cell division (Smith & Martin, 1973), which can be assessed directly, e.g., via light microscopy. These studies provided the experimental basis for the analysis of proliferation and created a need for comprehensive models describing growth dynamics of cell populations.

The first rigorous mathematical model for the growth dynamics of populations has been proposed by von Foerster (1959). The von Foerster equation,

$$\begin{aligned} &\frac{\partial}{\partial t} n(a|t) + \frac{\partial}{\partial a} n(a|t) = -\beta(a,t,\theta) n(a|t) \\ \text{with:} \quad &n(0|t) = \alpha(t,\theta), \qquad n(a|0) = n_0(a), \end{aligned} \tag{2.12}$$

is an one-dimensional partial differential equation describing the time dependent age distribution, $n(a|t) \in \mathbb{R}_+$, which is a number density function. Birth and death rate are denoted by $\alpha(a,t,\theta) : \mathbb{R}_+ \times \mathbb{R}_+ \times \mathbb{R}_+^{d_\theta} \rightarrow \mathbb{R}_+$ and $\beta(a,t,\theta) : \mathbb{R}_+ \times \mathbb{R}_+ \times \mathbb{R}_+^{d_\theta}$, respectively. The age distribution is a number density and the number of cells with age $a \in \Omega$ are given by $N(t) = \int_\Omega n(a|t) da$. As the von Foerster equation is a linear first order partial differential equation, its analytical solutions can be determined using the method of characteristics (Evans, 1998; Oldfield, 1966; Sinko & Streifer, 1967; Trucco, 1965).

To study other cellular properties besides age, i.e., protein abundance and cell volume, the von Foerster equation has been generalized to the so called population balance model (PBM) (Fredrickson *et al.*, 1967). The common population balance model is a first order integro-partial differential equation, allowing for the description of deterministic single cell dynamics and stochastic division times. These models are mainly used to examine the proliferation of microbial populations (Mantzaris, 2007; Tsuchiya *et al.*, 1966), e.g., in food industry, but further extensions also allow the study of T- and B-cell expansion (Banks *et al.*, 2010; De Boer *et al.*, 2006; Luzyanina *et al.*, 2009, 2007b), which is very important in immunology.

Unfortunately, for most of these more complex population balance models (PBM) the solution cannot be computed analytically. Instead, finite differences, finite volume, or finite elements discretization schemes are applied and the resulting ODE system is solved numerically. This need for numerical PDE solvers, which limits their state dimension to three, is the main drawback of PBMs and is probably the reason why they are almost exclusively used to study proliferation processes but not signal transduction events.

Chemical Master Equation and Fokker-Planck equation

To cope with signaling pathway models, the Chemical Master Equation (CME) and the Fokker-Planck equation (FPE) are employed (Gardiner, 2011; van Kampen, 2007). They provide a population level description for processes exhibiting stochastic dynamics.

Chemical Master Equation The CME is the statistical model of a cell population in which each individual follows a Markov jump process (as described in Section 2.1.1). The state variables of the CME are the probabilities $P(x|t) = P(X_t = x)$ that an individual cell X_t occupies a certain state $x \in \mathbb{N}^{d_s}$ (i.e., x_i molecules of species S_i) at time t. Hence, the state of the CME is a probability mass function in x.

The time evolution of the probability $P(x|t)$ is governed by the ODE (van Kampen, 2007),

$$\frac{d}{dt}P(x|t) = -P(x|t)\sum_{j=1}^{d_r} a_j(x,\theta) + \sum_{j=1}^{d_r} a_j(x - \nu_j, \theta)P(x - \nu_j|t), \tag{2.13}$$

which has to hold for all x, yielding a system of linear ODEs – the CME. Apparently, for most biological processes the number of reachable states x might be very large or even infinite rendering an analysis challenging.

To analyze the CME, different approaches have been introduced. For particular CMEs closed form solutions can be derived, e.g., for systems containing only monomolecular reactions with linear propensity functions (Jahnke & Huisinga, 2007), but for most CMEs a numerical approximation of the solution is necessary. Examples for numerical approaches are: the finite state projection (FSP) (Munsky & Khammash, 2006, 2008), employing an error-aware state truncation and numerical simulation; moment computation methods, relying on moment closure (Hespanha, 2007; Ruess *et al.*, 2011; Singh & Hespanha, 2011); and sample based approximation of the resulting distribution (El-Samad & Khammash, 2006). The latter method employes an ensemble model simulation and is therefore computationally demanding but also applicable for high-dimensional systems. The finite state projection on the other hand is very efficient in the case of low-dimensional processes but suffers the curse of dimensionality.

The CME has been employed to study, for example, gene regulation (Berg, 1978; El-Samad & Khammash, 2006; Paulsson, 2005) and cellular decision processes (Munsky & Khammash, 2006; Munsky *et al.*, 2009; Waldherr *et al.*, 2010). Similar to the simulation of Markov jump processes, also the simulation of the CME becomes intractable if the number of molecules is large. In these situations, the Fokker-Planck equation may be employed.

Fokker-Planck equation The FPE is the population level description of the chemical Langevin equation (Risken, 1996) and has as state variable the probability density $p(x|t)$. This probability density defines the probability for observing $X_t \in \Omega \subseteq \mathbb{R}^n_+$:

$$p(X_t \in \mathcal{X}) = \int_\Omega p(x|t)dx. \tag{2.14}$$

in which X_t is a realization of the chemical Langevin equation (2.6). The time evolution of $p(x|t)$ is governed by the PDE (Dargatz, 2010; Gardiner, 2011; Risken, 1996):

$$\frac{\partial}{\partial t}p(x|t) = -\sum_{i=1}^{d_s} \frac{\partial}{\partial x}\left[\mu_i(x,\theta)p(x|t)\right] + \frac{1}{2}\sum_{i_1=1}^{d_s}\sum_{i_2=1}^{d_s} \frac{\partial^2}{\partial x_{i_1}\partial x_{i_2}}\left[\Sigma_{i_1 i_2}(x,\theta)p(x|t)\right], \tag{2.15}$$

in which $\mu(x,\theta)$ is the drift vector and $\Sigma(x,\theta) = (\sigma(x,\theta))(\sigma(x,\theta))^\mathrm{T}$ diffusion matrix (see Section 2.1.1).

The FPE is the diffusion approximation of the CME. The key difference between both are the properties of x. For the CME the state x is integer-valued, whereas in case of the FPE x is real. This is why the distribution function for the former is a discrete (probability mass function) while it is continuous (probability density function) for the latter.

The FPE is widely used in biology as it allows – due the continuous state x – for the analysis of systems with large particle numbers. Applications of the FPE can be found in biochemical engineering (Mantzaris, 2007), molecular biology (Kepler & Elston, 2001), neurobiology (Galán *et al.*, 2007), as well as chemotaxis (Surulescu & Surulescu, 2010).

Unfortunately, as the method of characteristics cannot be applied (due to the diffusion), analytical solutions are available only in few cases (Risken, 1996). To solve the FPE, numerical PDE solvers, similar to those used for the population balance model, may be employed (Galán *et al.*, 2007). Therefore, dimensionality is also a problem. For higher-dimensional processes, i.e., $d_s \geq 4$, ensemble-based simulation methods (Kepler & Elston, 2001; Mantzaris, 2007) are used – like for the CME – to determine the characteristics of the population.

Besides the described PBM, CME and FPE, there are also generalizations available. A well-known example is the Cell Population Master Equation (Stamatakis & Zygourakis, 2010), which considers proliferation and signal transduction. These generalizations are biologically appealing as a variety of different effects may be considered, but due to their complexity an ensemble-based simulation is necessary resulting in high-computational effort and therefore limiting their application. Furthermore, none of these generalizations allows for the explicit consideration of cell-to-cell variability, e.g., in terms of parameter differences between individual cells.

2.3 Bayesian parameter estimation in a nutshell

In the previous section, models for single cells and cell populations were introduced. As discussed in Section 1.2.3, despite their different nature, all these models share one characteristic: their need for precise parameter values $\theta \in \mathbb{R}^{d_\theta}$. In this thesis, Bayesian methods are employed to infer the parameters θ from the available measurement data $\mathcal{D}$. A short introduction to Bayesian parameter estimation and uncertainty analysis is provided in this section.

2.3.1 Bayes' theorem and likelihood function

To estimate the parameters θ, the available (prior) information and the additional information, carried by the data $\mathcal{D}$ are merged. This merging of prior information and evidence (obtained by observing the process) is achieved using Bayes' theorem (MacKay, 2005),

$$\pi(\theta|\mathcal{D}) = \frac{\mathbb{P}(\mathcal{D}|\theta)\pi(\theta)}{\mathbb{P}(\mathcal{D})}. \tag{2.16}$$

Here, the prior distribution $\pi(\theta)$ and the posterior distribution $\pi(\theta|\mathcal{D})$ summarize the available information about θ before and after considering the data $\mathcal{D}$, respectively. Given $\pi(\theta)$, the posterior distribution $\pi(\theta|\mathcal{D})$ is computed using the likelihood $\mathbb{P}(\mathcal{D}|\theta)$ – also called conditional probability – and the marginal probability $\mathbb{P}(\mathcal{D})$. While $\mathbb{P}(\mathcal{D}) = \int_{\mathbb{R}_+^{d_\theta}} \mathbb{P}(\mathcal{D}|\theta)\pi(\theta)d\theta$ is simply a normalization constant, $\mathbb{P}(\mathcal{D}|\theta)$ is the probability density of observing the data $\mathcal{D}$ given the parameters θ.

The structure of Bayes' theorem is independent of the problem at hand, merely the likelihood function $\mathbb{P}(\mathcal{D}|\theta)$ varies. The likelihood function depends on the model of the dynamical system as well as on the model of the measurement process. For each combination a specific likelihood function has to be derived. In particular, for stochastic models the likelihood structures are involved and an evaluation is challenging (see Wilkinson (2010) and references therein).

Likelihood functions for deterministic processes are often easier, allowing, e.g., for factorization,

$$\mathbb{P}(\mathcal{D}|\theta) = \prod_{i=1}^{d_\mathcal{D}} \mathbb{P}(\mathcal{D}_i|\theta), \tag{2.17}$$

which enables the assessment of the information carried by the individual data points, $\mathcal{D} = \{\mathcal{D}_1, \mathcal{D}_2, \ldots, \mathcal{D}_{d_\mathcal{D}}\}$. To evaluate the likelihood of deterministic systems at a given point in parameter space, the model is simulated and the obtained output is compared to the measured data, accounting for the statistical properties of all measurement process. For stochastic systems an evaluation of the likelihood requires a marginalization over the possible stochastic realizations. Even for simple Markov jump processes and models based upon the chemical Langevin equation, this need for marginalization might prohibit the evaluation of the likelihood (Marjoram *et al.*, 2003; Wilkinson, 2010) or at least make it computationally demanding.

Only few problems allow for an explicit computation of the likelihood, and even fewer for an explicit formula of the posterior. Furthermore, the considered parameter spaces are often high-dimensional, rendering common gridding approaches useless as their computational effort grows exponentially with the number of parameters. To assess the properties of the posterior distribution $\pi(\theta|\mathcal{D})$, which encodes the knowledge about the parameters, sophisticated algorithms are required, such as nonlinear optimization or sampling.

2.3.2 Maximum a posteriori estimate, Bayesian confidence intervals and parameter dependencies

To analyze probability distributions, such as the posterior distribution $\pi(\theta|\mathcal{D})$, optimization-based and sample-based approaches are employed. A representative sample from $\pi(\theta|\mathcal{D})$ allows for the direct assessment of its statistical properties, e.g., mean and variances of the distribution, while optimizations can be used to determine the maximum a posteriori estimate.

Maximum a posteriori estimate

The maximum a posteriori estimate θ^* is the point in parameter space where the posterior probability $\pi(\theta|\mathcal{D})$ takes its maximum. Accordingly, θ^* is defined via the optimization problem

$$\theta^* = \arg\max_{\theta \in \mathbb{R}_+^{d_\theta}} \left[\pi(\theta|\mathcal{D}) = \frac{\mathbb{P}(\mathcal{D}|\theta)\pi(\theta)}{\mathbb{P}(\mathcal{D})} \right]. \tag{2.18}$$

This problem can be simplified by searching the maximum of the unnormalized posterior distribution, $\mathbb{P}(\mathcal{D}|\theta)\pi(\theta)$, instead of the maximum of the posterior distribution $\pi(\theta|\mathcal{D})$. This avoids the computation of the normalization constant $\mathbb{P}(\mathcal{D})$, which is known to be a demanding problem (Kramer *et al.*, 2010). Furthermore, a log-transformation of the objective ensures a further improvement due to better numerical properties. Taken together, these two modifications yield the maximization problem,

$$\theta^* = \arg\max_{\theta \in \mathbb{R}_+^{d_\theta}} \left[\log(\mathbb{P}(\mathcal{D}|\theta)) + \log(\pi(\theta)) \right], \tag{2.19}$$

which is equivalent to (2.18). An additional log-transformation of the parameters, $\theta = \exp(\bar{\theta})$, and estimation of the exponents $\bar{\theta} \in \mathbb{R}^{d_\theta}$ also offers the potential of improved efficiency. The advantage is that all entries of the transformed parameter $\bar{\theta}$ are of the same order of magnitude.

Independent of the precise formulation, the maximization problem is in general nonlinear and non-concave. To solve (2.19), sophisticated optimization schemes are required. Common tools employ multiple shooting (Biegler, 2007; Bock & Plitt, 1984; Peifer & Timmer, 2007), evolutionary algorithms (Bock & Plitt, 1984), particle swarm optimization (Vaz & Vicente, 2007), and pattern search (Vaz & Vicente, 2007). Banga (2008) and Weise (2009) provide comprehensive surveys of local and global optimization procedures.

The maximum a posteriori estimate is closely related to the maximum likelihood estimate. It provides a point estimate of the parameter values. As the maximum likelihood objective is augmented by the prior information, the maximum a posteriori estimate may be interpreted as a regularized – and hence stabilized – version of the maximum likelihood estimate. If a single parameter would have to be selected for the model, the maximum a posteriori estimate would be the parameter of choice.

Unfortunately, the maximum a posteriori estimate might not be unique, if the model is not identifiable, and often depend strongly on the measurement noise. Thus, the computed value for θ^* might be unreliable. This is suboptimal, as conclusions draw using the model might rely heavily on θ^*. Different values for θ^* may result in different predictions and establish wrong expectations. To evaluate the model uncertainties, we have to go beyond point estimates, like θ^*, and study the global properties of $\pi(\theta|\mathcal{D})$.

Bayesian confidence intervals

The global properties of a probability distribution can be studied using a representative sample, $\mathcal{S} = \{\theta^k\}_{k=1}^{d_S} = \{\theta^1, \theta^2, \ldots, \theta^{d_S}\}$. The analysis of a sample $\mathcal{S}$ reveals parameter uncertainties as well as correlations.

In Bayesian statistics, the parameter uncertainty is studied via the Bayesian confidence intervals (Chen & Shao, 1999), also known as credible intervals. A $100(1-\alpha)\%$ Bayesian confidence interval for parameter θ_j is any set $[\theta]_j$ containing $100(1-\alpha)$ percent of the probability mass. Thus, it has to hold that

$$\int_{\{\theta \in \mathbb{R}_+^{d_\theta} | \theta_j \in [\theta]_j\}} \pi(\theta|\mathcal{D}) d\theta = 1 - \alpha. \tag{2.20}$$

Apparently, (2.20) does not define $[\theta]_j$ uniquely. Further specifications are necessary, employing, e.g., the posterior density (Chen & Shao, 1999).

In this work, Bayesian confidence intervals are defined via the percentiles of sample $\mathcal{S}$ (DiCiccio & Efron, 1996; Joshi *et al.*, 2006). The 100α-th percentile of a random variable θ_j is the value $\theta_j^{(\alpha)}$ below which $100\alpha\%$ of the observations fall. Accordingly, the $100(1-\alpha)$-th percentile interval of θ_j is defined as $[\theta_j^{(\alpha/2)}, \theta_j^{(1-\alpha/2)}]$. The Bayesian confidence intervals $[\theta]_j$ are frequently defined as the 95-th percentile interval. Thus, the parameter is contained in $[\theta]_j^{0.95} = [\theta_j^{(0.025)}, \theta_j^{(0.975)}]$ with a probability of 95% given the measurement data and the prior knowledge.

Bayesian confidence intervals provide a measure for the uncertainty of the parameters. Wide intervals indicate a large uncertainty, while narrow intervals indicate small uncertainties. For a multi-dimensional analysis, a generalization of Bayesian confidence intervals to Bayesian confidence regions may be employed.

To assess the prediction power of models, also prediction uncertainties have to be studied. As model predictions are often time-dependent, e.g., the concentration of a protein, time-dependent Bayesian confidence intervals are required. These are easily obtained by computing the Bayesian confidence intervals at the individual time point of interest.

Parameter dependencies

Beyond the analysis of confidence intervals, we are interested in the analysis of co-dependency of parameters. Frequently, it is observed that sums and products of individual parameters can be estimated precisely, although the uncertainty of individual parameters is large. Such co-dependencies can be assessed using a principle components analysis (Jolliffe, 2002) or a correlation analysis (e.g., Pearson correlation or Spearman correlation) (Rodgers & Nicewander, 1988) applied to the sample. Alternatively, the mutual information or the maximum information coefficient can be computed (Reshef *et al.*, 2011) providing better measures in case of complex co-dependencies. Knowledge about principle components and correlations allows the unraveling of stiff and sloppy directions in parameter space (Apgar *et al.*, 2010; Gutenkunst *et al.*, 2007) and may be used to reduce the number of unknown parameters (Balsa-Canto *et al.*, 2010).

2.3.3 Markov chain Monte Carlo sampling

For the analysis of parameter and prediction uncertainties, a representative sample of the posterior distribution $\pi(\theta|\mathcal{D})$ is required. The most common methods to generate such as

Algorithm 2.1 Pseudocode for Markov chain Monte Carlo sampling.

Require: data $\mathcal{D}$, prior distribution $\pi(\theta)$, likelihood function $\mathbb{P}(\mathcal{D}|\theta)$, chain length d_S.
Require: initial parameter vector θ^0, prior probability density $\pi(\theta^0)$, likelihood $\mathbb{P}(\mathcal{D}|\theta^0)$.
  for $k = 1$ to d_S **do**
    Propose θ^k by sampling from the proposal density, $\theta^k \sim \mathbb{Q}(\theta^k, \theta^{k-1})$.
    Evaluate likelihood $\mathbb{P}(\mathcal{D}|\theta^k)$ and prior probability density $\pi(\theta^k)$.
    Draw random number r from uniform distribution, $r \sim \mathbb{U}_{[0,1]}(r)$.
    if $r < p_a = \min\left\{1, \frac{\mathbb{P}(\mathcal{D}|\theta^k)}{\mathbb{P}(\mathcal{D}|\theta^{k-1})} \frac{\pi(\theta^k)}{\pi(\theta^{k-1})} \frac{\mathbb{Q}(\theta^{k-1},\theta^k)}{\mathbb{Q}(\theta^k,\theta^{k-1})}\right\}$ **then**
      Accept proposed parameter vector θ^k.
    else
      Reject proposed parameter vector θ^k, $\theta^k = \theta^{k-1}$.
    end if
  end for

sample are: direct sampling, rejection sampling and Markov chain Monte Carlo (MCMC) sampling (MacKay, 2005). While direct sampling is only applicable if a closed form solution of the posterior is available, which is unusual for parameter estimation problems, rejection sampling is known to perform poorly for high-dimensional problems. For these reasons, MCMC sampling is employed in this thesis.

Markov chain Monte Carlo methods generate a chain of parameters, $\theta^1, \theta^2, \ldots, \theta^{d_S}$, by exploring the posterior distribution $\pi(\theta|\mathcal{D})$. This exploration employs a two-step procedure, which is shared among most MCMC methods, among others Metropolis-Hastings sampling (Chib & Greenberg, 1995), Gibbs sampling, and slice sampling (MacKay, 2005).

In the first step – the proposal step –, a point θ^k in parameter space is proposed. Therefore, we draw from the proposal density,

$$\theta^k \sim \mathbb{Q}(\theta^k, \theta^{k-1}). \tag{2.21}$$

The proposal density $\mathbb{Q}(\theta^k, \theta^{k-1})$ depends on the previous chain member, θ^{k-1}, and might be any distribution from which we can sample, such as a Gaussian centered at θ^{k-1}. In the second step – the acceptance/rejection step –, the proposed parameter vector θ^k is accepted or rejected. The probability to accept θ^k is

$$p_a = \min\left\{1, \frac{\mathbb{P}(\mathcal{D}|\theta^k)}{\mathbb{P}(\mathcal{D}|\theta^{k-1})} \frac{\pi(\theta^k)}{\pi(\theta^{k-1})} \frac{\mathbb{Q}(\theta^{k-1}, \theta^k)}{\mathbb{Q}(\theta^k, \theta^{k-1})}\right\}. \tag{2.22}$$

Hence, it is determined by the ratio of the posterior probability density, $\frac{\pi(\theta^k|\mathcal{D})}{\pi(\theta^{k-1}|\mathcal{D})} = \frac{\mathbb{P}(\mathcal{D}|\theta^k)}{\mathbb{P}(\mathcal{D}|\theta^{k-1})} \frac{\pi(\theta^k)}{\pi(\theta^{k-1})}$, and the proposal density. For symmetric proposals, $\mathbb{Q}(\theta^{k-1}, \theta^k) = \mathbb{Q}(\theta^k, \theta^{k-1})$, the last fraction is equal to one while for asymmetric proposals it is different from one to ensure reversibility. If the proposed parameter vector is rejected, the previous parameter vector is restored, $\theta^k = \theta^{k-1}$, and the counter is updated, $k = k + 1$. The pseudocode of the MCMC sampling is depicted in Algorithm 2.1.

It can be verified that the stationary distribution of the random chain, $\theta^1, \theta^2, \ldots, \theta^{d_S}$, generated using the above two-step procedure, is the posterior distribution. Thus, this method provides a representative sample from $\pi(\theta|\mathcal{D})$ (Chib & Greenberg, 1995; MacKay, 2005), if the sample is large enough ($d_S \gg 1$). The size of the sample is crucial, as, unlike direct and rejection sampling, MCMC sampling yields correlated samples. The k-th member of the chain, θ^k, depends on the $(k-1)$-th chain member, θ^{k-1}. Thus, the algorithm possesses the Markov property, which is why such sampling procedures are called Markov chain Monte Carlo methods. To check the convergence of MCMC sampling, different methods are known (Brooks & Roberts, 1998), e.g., Geweke's spectral density diagnostics.

While the proposal density $\mathbb{Q}(\theta^k, \theta^{k-1})$ might be any distribution, the efficiency of the sampling, in particular the correlation length of the sample, strongly depends on the precise choice. To achieve good sampling performance, schemes for the online adaptation of the proposal density $\mathbb{Q}(\theta^k, \theta^{k-1})$ have been developed (Haario *et al.*, 2006, 2001). These schemes optimize the parameters of the proposal density, e.g., the covariance of a Gaussian proposal density, employing the knowledge about the posterior distribution encoded in $\theta^1, \theta^2, \ldots, \theta^{k-1}$. Such adaptive MCMC sampling can be backed up by a delay rejection (Green & Mira, 2001; Haario *et al.*, 2006), which can improve the performance in case of locally badly calibrated proposals. Further improvement is achieved by applying sequential methods (Doucet *et al.*, 2000), which bridge from the prior to the posterior distribution via intermediate distributions.

As for stochastic processes the evaluation of the likelihood may be intractable, likelihood-free (Marjoram *et al.*, 2003; Sisson *et al.*, 2007) sampling approaches and approximate Bayesian computation (Beaumont *et al.*, 2002; Toni & Stumpf, 2010; Toni *et al.*, 2009) have been introduced. These methods employ the idea of MCMC sampling, but bypass the evaluation of the likelihood. Instead, as single realization of the stochastic process is generated – this is often efficient –, and compared to the measurement data. It turns out that the stationary distribution of Markov chains obtained using likelihood-free MCMC sampling coincides with the posterior distribution (Marjoram *et al.*, 2003). For approximate Bayesian computation this is not true, but the level of acceptable error can be selected (Beaumont *et al.*, 2002).

This short overview already shows that a multitude of differential MCMC schemes exist. In this work, we employ the adaptive Metropolis-Hastings scheme with delayed rejection introduced in (Haario *et al.*, 2006). This MCMC algorithm is implemented in a `MATLAB` toolbox, available at `http://www.helsinki.fi/~mjlaine/mcmc/`. The algorithm of this toolbox has been proven efficient for a variety of examples.

Summing up, in this section, we introduced the fundamentals of Bayesian estimation, Bayesian statistics, and MCMC sampling. For a comprehensive overview, the overall Bayesian parameter estimation procedure is illustrated in Figure 2.3. Apparently, we only scratched the surface of Bayesian estimation. For an in-depth discussion we refer to (MacKay, 2005; Wilkinson, 2007, 2010) and references therein.

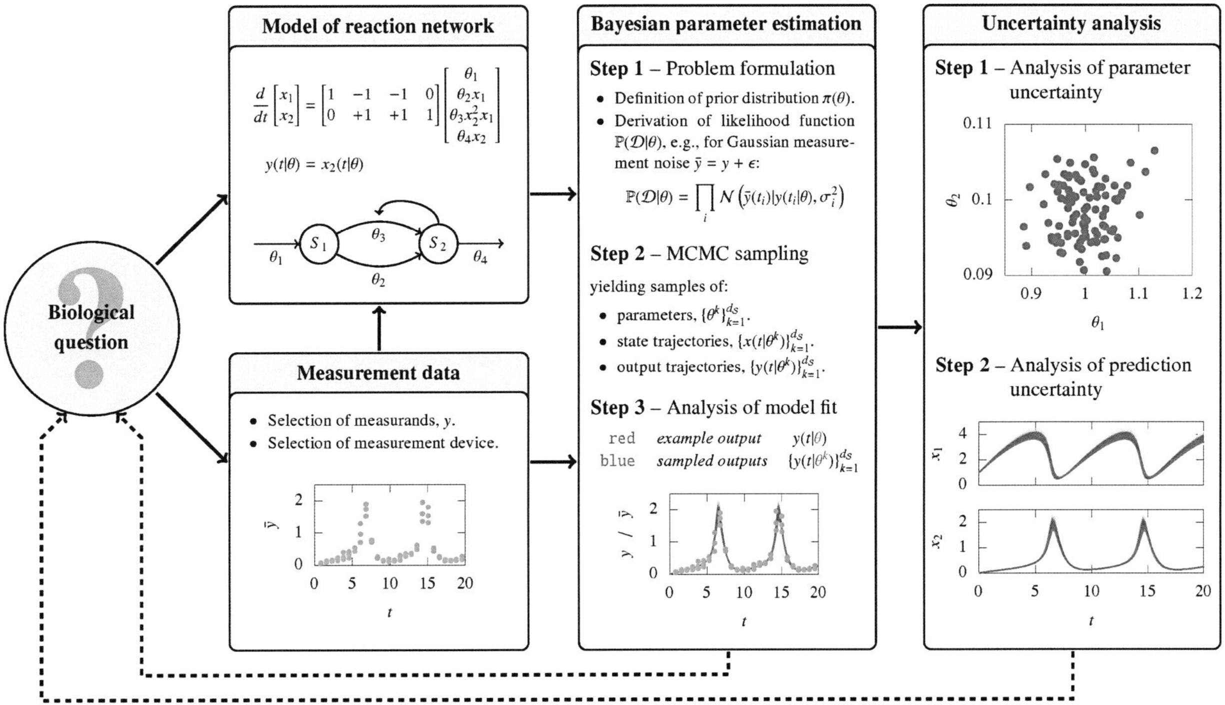

Figure 2.3: Illustration of parameter estimation workflow, including: data collection, model development, parameter estimation, and uncertainty analysis. Also feedback from parameter estimation and uncertainty analysis to the biological question are indicated (---). Such feedback might be the rejection or validation of (model) hypothesis, prediction about unmeasured quantities, as well as candidates for future experiments. While the available data and the models strongly depend on the biological question, the workflow is identical.

3 Signal transduction in heterogeneous cell populations

In this chapter, we study signal transduction in heterogeneous cell populations. Therefore, we provide in Section 3.1 a formal introduction to the considered process class and the available single cell snapshot data that prevail in practice. Afterwards, we derive in Section 3.2 a population level model for signal transduction in cell populations exhibiting stochastic and deterministic cell-to-cell variability, employing an augmented Fokker-Planck equation. In Section 3.3, a Bayesian approach for parameter estimation and uncertainty analysis is proposed. This approach is based upon a subdivision of the estimation problem into a pre-estimation phase, during which a simple parametric form of the likelihood function is computed, and an estimation phase. Modeling and parameter estimation are applied to a model of apoptotic signaling in Section 3.4.

The individual parts of this chapter are based upon (Hasenauer *et al.*, 2011c,d, 2010a,b).

3.1 Introduction and problem statement

3.1.1 Biological background

One of the main reasons for cell-to-cell variability are differences in cellular information processing. Even cells in clonal populations may respond differentially to the same stimulus, inducing phenotypic heterogeneity. This has been observed for cell fate decision making in stem cells (Glauche *et al.*, 2010; Schroeder, 2011;Schittler *et al.*, 2010) and in cancer cells (Eissing *et al.*, 2009; Paszek *et al.*, 2010; Singh *et al.*, 2010; Spencer *et al.*, 2009; Yang *et al.*, 2010), the infection of cells by viruses (Rosenfeld *et al.*, 2005; Snijder & Pelkmans, 2011), the sensitization of neurons (Andres *et al.*, 2010; Andres *et al.*, 2012), the food source selection of microbes (Mantzaris, 2007; Munsky *et al.*, 2009; Song *et al.*, 2010), and for many other processes (see, e.g., (Henson, 2003)).

There are two sources for phenotypic heterogeneity: intrinsic noise and extrinsic noise. Following the definition provided by Swain *et al.* (2002), intrinsic noise refers to the inherent stochasticity of the biochemical process. Considering one chemical species in the cell, this stochasticity arises especially from stochastic gene expression (Eldar & Elowitz, 2010; Elowitz *et al.*, 2002; Lestas *et al.*, 2010; Longo & Hasty, 2006; Rosenfeld *et al.*, 2005). Extrinsic variability on the other hand refers to variations in the abundance of other chemical species, e.g., transcription factors and ribosomes, and variations in biochemical conditions, such as the reaction volume (this is the volume of the cell or of individual organelles). Also chemical species subject to intrinsic noise may provide a source of extrinsic noise for other chemical species with which it interacts (Huh & Paulsson, 2011; Swain *et al.*, 2002).

While intrinsic noise is purely random, extrinsic noise can also be (partially) regulated (Snijder & Pelkmans, 2011). The most common examples are epigenetic differences (Avery,

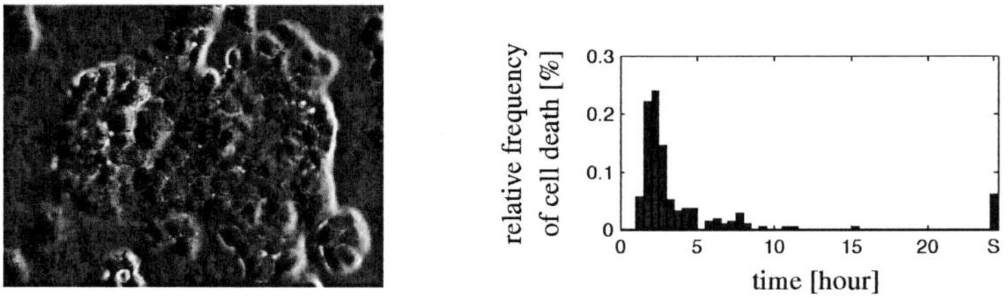

(a) Population of NCI H460 cells after stimulation with 100 ng/ml TRAIL.

(b) Relative frequency of cell death and percentage of surviving cells (S).

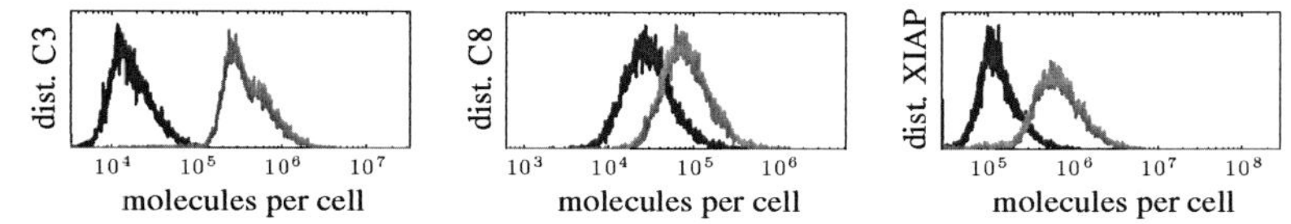

(c) Abundance of the key regulator caspase 3, caspase 8, and XIAP in NCI H460 cells. The dark gray histograms represent unlabeled control cells, and provide a measure for the autofluorescence, while the light gray histograms represent the measured proteins distributions.

Figure 3.1: Illustration of heterogeneity in a population of NCI-H460 cancer cell. A population of NCI-H460 cancer cells (a) is stimulated with 100 ng/ml Tumor Necrosis Factor Related Apoptosis Inducing Ligand (TRAIL) and observed using time-lapse microscopy. The distribution of death times (b) reveals that individual cells undergo apoptosis at greatly different time instances. Furthermore, some of the cells survive. This different phenotypes may arise from variations in the abundance of key proteins (c), caused by intrinsic and extrinsic noise.[1]

2006), differences in the cell cycle stage (Huh & Paulsson, 2011; Swain *et al.*, 2002), unequal partitioning of cellular material at cell division (Huh & Paulsson, 2011; Mantzaris, 2007) – this is particularly relevant for different yeast strains (Gyllenberg, 1986) –, and differences in the microenvironment of cells (e.g., the local cell density) (Snijder & Pelkmans, 2011). Regulated cell-to-cell variability also gives rise to population heterogeneity, but the decision process determining the phenotypes is not longer purely stochastic.

First studies on the impact of the different sources have been carried out in (Hilfinger & Paulsson, 2011; Huh & Paulsson, 2011; Lestas *et al.*, 2010; Mantzaris, 2007; Spencer *et al.*, 2009; Stamatakis & Zygourakis, 2010; Swain *et al.*, 2002). It has been shown that the noise sources cannot be suppressed completely, and that the benefits of feedback are limited (Huh & Paulsson, 2011; Lestas *et al.*, 2010). Furthermore, these studies revealed – for different models – that intrinsic noise, extrinsic noise, and a combination of both might result in complex population dynamics, such as bimodal distributions and different phenotypes (Mantzaris, 2007; Spencer *et al.*, 2009). In particular for apoptotic signaling this is well established and illustrated in Figure 3.1. Here, two extreme phenotypes – surviving and dying cells –, are observed due to the differential abundance of pro- and antiapoptotic factors.

[1]The illustrated experimental data have been provided by Malgorzata Doszczak and Peter Scheurich (Institute for Cell Biology and Immunology, University of Stuttgart, Germany).

3.1.2 Modeling of single cells in a heterogeneous population

In this thesis, we account for stochastic and deterministic sources of cell-to-cell variability. Individual cells are modeled by a parametric chemical Langevin equation. As discussed in Section 2.1.1, the chemical Langevin equation belongs to the class of stochastic differential equations. This modeling framework allows to describe metabolic networks as well as signal transduction pathways, as long as spatial effects can be neglected. Furthermore, the stochastic differential equations comprise the ordinary differential equations.

Mathematically, the dynamic behavior of each cell is determined by

$$dX_t = \mu(X_t, \theta, t)dt + \sigma(X_t, \theta, t)dW_t \tag{3.1}$$

with state variables $X_t \in \mathbb{R}_+^{dx}$, initial state $X_0 \in \mathbb{R}_+^{dx}$, Wiener processes $W_t \in \mathbb{R}^{dw}$, and the parameters $\theta^{(i)} \in \mathbb{R}_+^{d_\theta}$. The locally Lipschitz vector fields $\mu : \mathbb{R}_+^{dx} \times \mathbb{R}_+^{d_\theta} \times \mathbb{R}_+ \to \mathbb{R}_+^{dx}$ and $\sigma : \mathbb{R}_+^{dx} \times \mathbb{R}_+^{d_\theta} \times \mathbb{R}_+ \to \mathbb{R}_+^{dx \times dw}$ describing the deterministic and the stochastic evolution of a single cell, respectively (see Section 2.1.1). To enable the consideration of external inputs and dynamic perturbations, we allow for an explicit time dependence of μ and σ. Parameters θ can be: kinetic constants, e.g., reaction rate constants or binding affinities; abundances of, for instance, transcription factors, activators or inhibitors; but also overall cell properties, such as the cycle stage or the cell volume.

The source of deterministic (or regulated) cell-to-cell variability are differences in the parameters of individual cells. Within the cell population, the parameters are distributed according to a probability density function $p_\theta(\theta) : \mathbb{R}_+^{d_\theta} \to \mathbb{R}_+$. This means that the parameter vector of an individual cell, θ, has probability

$$\text{Prob}(\theta \in \Omega) = \int_\Omega p_\theta(\theta)d\theta \tag{3.2}$$

of being contained in the parameter set Ω.

The probability density $p_\theta(\theta)$ might be multivariate, allowing for co-dependencies among parameters. In addition, the initial condition of individual cells, X_0, might partially depend on the parameters. To account for this, we define the combined vector of parameters and initial conditions, $Z_0 = [X_0^\mathrm{T}, \theta^\mathrm{T}]^\mathrm{T} \in \mathbb{R}_+^{dz}$ with $d_Z = d_X + d_\theta$, and the corresponding joint probability density of parameters and initial conditions, $p_0(z_0)$. Accordingly,

$$\text{Prob}(Z_0 \in \Omega) = \int_\Omega p_0(z_0)dz_0 \tag{3.3}$$

is the probability that $Z_0 \in \Omega$. The parameter density can be determined from $p_0(z_0)$ by marginalization over the first d_X dimensions of $p_0(z_0)$, $p_\theta(\theta) = \int_0^\infty p_0([x_0^\mathrm{T}, \theta^\mathrm{T}]^\mathrm{T})dx_0$.

By allowing for stochastic cell-to-cell variability (Wiener process) and deterministic cell-to-cell variability (variation of parameters), we govern a wide range of biological processes involving cell populations. Merely cell-cell communication in terms of juxtacrine, paracrine and endocrine signaling, and indirect cell-cell interactions, e.g., via conditioning of the growth medium, are disregarded. These simplifying assumptions are indeed fulfilled for many *in vitro* lab experiments, where the response of the individual cells is predominantly influenced by external stimuli.

A mechanistic description (3.1) of pathways can often be derived from the literature. Therefore, the main problem is that the parameter distribution, $p_\theta(\theta)$, and the joint distribution of parameters and initial conditions, $p_0(z_0)$, are in general unknown and have to be determined.

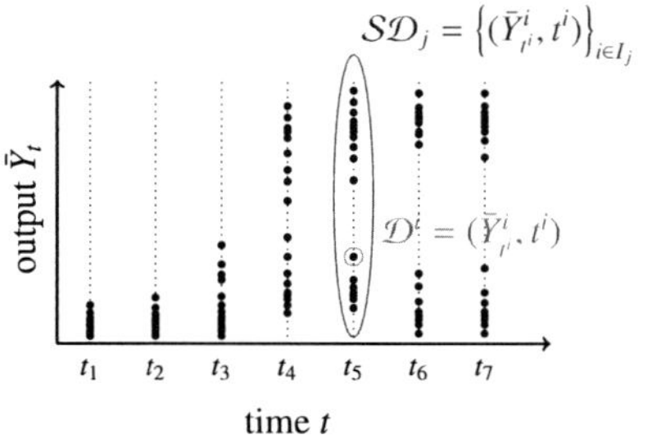

Figure 3.2: Detailed illustration of an exemplary population snapshot data of a heterogeneous cell population. The single cell measurement (•) is denoted by $\mathcal{D}^i$ and the snapshot at a particular time instance t_j is denoted by $\mathcal{SD}_j$. The collection of all data is referred to as $\mathcal{D}$.

3.1.3 Measurement data

In this thesis, we consider single cell snapshot data (see Figure 3.2). As outlined in Section 1.2.1, such data can be collected using high-throughput fluorescence flow cytometry (Herzenberg *et al.*, 2006) or microscopy (Andres *et al.*, 2010). The corresponding data provide information about individual cells but also about the cell population.

Single cell data

The single cell data collected via flow cytometry and microscopy are typically mRNA, protein, or metabolite concentrations, but also measurements of protein activities are feasible. Mathematically, single cell measurements are described by

$$Y_t = \gamma(X_t, \theta, t), \tag{3.4}$$

in which the function $\gamma : \mathbb{R}_+^{d_X} \times \mathbb{R}_+^{d_\theta} \times \mathbb{R}_+ \to \mathbb{R}_+^{d_Y}$ is continuously differentiable. The function γ depends on the experimental setup, e.g., the used antibodies or labels. If the concentration of the k-th biochemical species, $X_{k,t}$, is observed via flow cytometry using a fluorescent label, we would have $Y_t = \gamma(X_t, \theta, t) = \theta_j X_{k,t}$, where θ_j is a proportionality factor associated to the labeling efficiency and the fluorescence of the antibody (Weber *et al.*, 2011).

Like all measurement devices, also high-throughput fluorescence measurements are subject to measurements errors. The noise-corrupted output, $\bar{Y}_t$, can be described as single realization from a probability density,

$$\bar{Y}_t \sim p(\bar{Y}_t | Y_t), \tag{3.5}$$

in which Y_t is the actual output from (3.4). The precise structure and magnitude of the measurement error is problem dependent and influenced by many factors, among others the sensitivity and specificity of antibodies (Andres *et al.*, 2012). For experimental setups in biology, the measurement noise often has absolute and relative components (Herzenberg *et al.*, 2006; Kreutz *et al.*, 2007). While the statistics of the former do not depend on the state of the system, the latter are state dependent and increase as Y_t increases. In Section 3.4, a typical noise model will be introduced.

For the theoretical part of this chapter, we do not assume any specific noise distribution.

Single cell snapshot data

Measurement devices collecting single cell snapshot data observe populations of cells at different time points t_j, $j = 1, \ldots, d_{\mathcal{SD}}$. Each observation is called a snapshot, as it provides information about the single cells for this particular time instance (see Figure 3.2). Due to the experimental setup, cell populations analyzed at different time instances are different ones, e.g., from different culture wells. Therefore, the number of cells contained in the j-th snapshot, $\bar{N}_{t_j}$, may differ for $j = 1, \ldots, d_{\mathcal{SD}}$.

The single cells in the individual snapshots are referenced through index sets. The j-th snapshot contains all cells from the index set $I_j = \left\{\sum_{k=0}^{j-1} \bar{N}_{t_k} + 1, \ldots, \sum_{k=0}^{j} \bar{N}_{t_k}\right\}$, with $\bar{N}_{t_0} = 0$. The data point for the cell with index i is denoted by

$$\mathcal{D}^i = \left(\bar{Y}^i_{t^i}, t^i\right), \tag{3.6}$$

in which t^i is the time at which the measurement was performed, and $\bar{Y}^i_{t^i}$ is the measured, noise-distorted output as defined above. If cell i has been measured as part of the j-th snapshot, then $t^i = t_j$. The j-th snapshot is the set of all data points $\mathcal{D}^i$ with $i \in I_j$, as depicted in Figure 3.2:

$$\mathcal{SD}_j = \bigcup_{i \in I_j} \mathcal{D}^i = \left\{\left(\bar{Y}^i_{t^i}, t^i\right)\right\}_{i \in I_j}. \tag{3.7}$$

To ensure a compact notation, also the collection of all data is introduced,

$$\mathcal{D} = \bigcup_{j=1}^{d_{\mathcal{SD}}} \mathcal{SD}_j = \left\{\left(\bar{Y}^i_{t^i}, t^i\right)\right\}_{i=1}^{\bar{N}}, \tag{3.8}$$

in which $\bar{N} = \bar{N}_{t_1} + \bar{N}_{t_2} + \ldots + \bar{N}_{t_{d_{\mathcal{SD}}}}$ is the total number of measured cells.

We emphasize that experimental setups are considered in which cells are not tracked over time. These setups are very common in studies on the population scale. Again, classical examples for measurement techniques yielding such data are flow cytometric analysis and cytometric fluorescence microscopy. As the population is well mixed when the measurement is performed and no cell is measured more than once, the individual single cell measurements, $\mathcal{D}^i$, are independent. This independence of $\mathcal{D}^{i_1}$ and $\mathcal{D}^{i_2}$ (respectively $\bar{Y}^{i_1}_{t^{i_1}}$ and $\bar{Y}^{i_2}_{t^{i_2}}$), $i_1 \neq i_2$, holds if both cells are measured during one snapshot ($t^{i_1} = t^{i_2}$) as well as if the cells are measured within different snapshots ($t^{i_1} \neq t^{i_2}$).

3.1.4 Problem formulation

Given the above setup, we face several challenges. First of all, the populations to be studied in standard lab experiments are on the order of $10^3 - 10^6$ individual cells. A direct simulation of such populations on an individual basis is intractable. This renders a density-based population analysis necessary.

Problem 3.1. (Modeling of heterogeneous cell population) *Given the model of the single cell dynamics* (3.1)*, the output mapping* (3.4)*, and the initial density,* $p_0(z_0)$*, describe the time evolution of the cell population density.*

Such density-based descriptions are available in case of purely stochastic dynamics, as discussed in Section 2.2.2. A combined population model for stochastic single cell dynamics (modeled by SDEs) and heterogeneity in parameters is not yet available in the literature to the best knowledge of the author of this thesis.

In addition, the parametric heterogeneity can in general not be assessed experimentally. Therefore, this source of cell-to-cell variability has to be reconstructed using the available single cell snapshot data. An

Problem 3.2. (Estimation and uncertainty analysis for parameter distribution) *Given the measurement data $\mathcal{D}$, the noise model* (3.5)*, and a model for the population dynamics, estimate the joint density of parameter and initial conditions, $p_0([x_0^{\mathrm{T}}, \theta^{\mathrm{T}}]^{\mathrm{T}})$, and its uncertainty.*

While both, the forward problem (Problem 3.1) and the backward problem are important, the inverse problem (Problem 3.2) is far more challenging. In the previous setup this is particularly true as we have to infer a multi-dimensional function.

3.2 Modeling of signal transduction in heterogeneous cell populations

In this section, we approach Problem 3.1 and derive a partial differential equation model governing the population dynamics. Following that, an efficient simulation scheme is presented.

3.2.1 Augmented Fokker-Planck equation for heterogeneous populations

A single cell model describes the state of an individual cell and its dynamics. Accordingly, a population model describes the state of the population and its dynamics. As populations consist of individual cells, a plausible choice for the state of the population is the distribution (frequency) of single cell states. Accordingly, if the single cell state is continuous, the population model should provide the probability density of observing a certain single cell state. To ensure this, the population model has to respect the dynamics of the individual cells, thereby providing a mechanistic population description.

To derive the population model, we define the augmented single cell model:

Definition 3.1. *The augmented single cell dynamics for the chemical Langevin equation* (3.1) *and output mapping* (3.4) *is defined by:*

$$\begin{aligned} dZ_t &= \tilde{\mu}(Z_t, t)dt + \tilde{\sigma}(Z_t, t)dW_t \\ Y_t &= \tilde{\gamma}(Z_t, t), \end{aligned} \tag{3.9}$$

with state variable $Z_t = [(Z_t^{(1)})^{\mathrm{T}}, (Z_t^{(2)})^{\mathrm{T}}]^{\mathrm{T}} \in \mathbb{R}_+^{d_Z}$ and initial state $Z_0 = [X_0^{\mathrm{T}}, \theta^{\mathrm{T}}]^{\mathrm{T}}$. The augmented drift function $\tilde{\mu}(Z_t, t) : \mathbb{R}_+^{d_Z} \times \mathbb{R}_+ \to \mathbb{R}_+^{d_Z}$, the augmented diffusion function $\tilde{\sigma}(Z_t, t) : \mathbb{R}_+^{d_Z} \times \mathbb{R}_+ \to \mathbb{R}_+^{d_Z \times d_W}$, and the augmented output map $\tilde{\gamma}(Z_t, t) : \mathbb{R}_+^{d_Z} \times \mathbb{R}_+ \to \mathbb{R}_+^{d_Y}$ are defined as:

$$\tilde{\mu}(Z_t) := \begin{pmatrix} \mu(Z_t^{(1)}, Z_t^{(2)}, t) \\ 0 \end{pmatrix}, \; \tilde{\sigma}(Z_t) := \begin{pmatrix} \sigma(Z_t^{(1)}, Z_t^{(2)}, t) \\ 0 \end{pmatrix}, \; \textit{and } \tilde{\gamma}(Z_t) := \gamma(Z_t^{(1)}, Z_t^{(2)}). \tag{3.10}$$

This augmented single cell model is a parameter-free stochastic differential equation. Its state is a composition of the actual state of the single cell, $Z_t^{(1)} = X_t$, and the parameters, $Z_t^{(2)} = \theta$. Thus, Z_t contains all information about the single cell, while X_t does not contain information about θ. In the augmented system, different parameter values are encoded in differential initial conditions, $Z_0^{(2)} = \theta$.

Employing (3.9), a model for the population density, $p(z|t, p_0)$, is derived. The population density is a probability density function. It provides the probability of drawing at random a cell from the population with state vector $Z_t \in \Omega$ at time t,

$$\mathrm{Prob}(Z_t \in \Omega) = \int_{\Omega} p(z|t, p_0) dz. \tag{3.11}$$

The change of $p(z|t, p_0)$ depends on the augmented single cell dynamics:

Theorem 3.2. *Let* (3.1) *be the stochastic dynamics of the individual cells and* $p_0([x_0^{\mathrm{T}}, \theta^{\mathrm{T}}]^{\mathrm{T}})$ *the joint probability density of states and parameters at time zero. Then, the probability density of the augmented state,* $p(z|t, p_0)$*, evolves according to*

$$\frac{\partial}{\partial t} p(z|t, p_0) = -\sum_{i=1}^{d_Z} \frac{\partial}{\partial z_i} \left[\tilde{\mu}_i(z,t) p(z|t, p_0)\right] + \frac{1}{2} \sum_{i=1}^{d_Z} \sum_{j=1}^{d_Z} \frac{\partial^2}{\partial z_i \partial z_j} \left[\tilde{\Sigma}_{ij}(z,t) p(z|t, p_0)\right], \tag{3.12}$$

with initial condition $p(z|0, p_0) \equiv p_0(z)$ *and diffusion term* $\tilde{\Sigma}(z,t) = (\tilde{\sigma}(z,t))(\tilde{\sigma}(z,t))^{\mathrm{T}}$.

Proof. To prove Theorem 3.2, note that the augmented single cell model is a common SDE. The probability density of the state of a SDE is known to be governed by the Fokker-Planck equation (Dargatz, 2010; Gardiner, 2011; Risken, 1996). The Fokker-Planck equation for the augmented single cell model – in the following called augmented Fokker-Planck equation – is given by (3.12), which concludes the proof. □

The augmented Fokker-Planck equation (3.12) is linear and its solution exists for sufficiently smooth $\tilde{\mu}(z,t)$, $\tilde{\Sigma}(z,t)$, and smooth initial conditions $p_0(z_0)$ (Evans, 1998). Due to its linearity, we can prove easily that the superposition principle holds:

Proposition 3.3. *Let* $p(z|t, p_{0,1})$ *and* $p(z|t, p_{0,2})$ *be the solution of the augmented Fokker-Planck equation* (3.12) *for initial probability densities,* $p_{0,1}(z_0)$ *and* $p_{0,2}(z_0)$*, respectively. Then, the solution of* (3.12) *for*

$$p_0(z_0) = \varphi p_{0,1}(z_0) + (1-\varphi) p_{0,2}(z_0), \qquad 0 \leq \varphi \leq 1, \tag{3.13}$$

is

$$p(z|t, \varphi p_{0,1} + (1-\varphi) p_{0,2}) = \varphi p(z|t, p_{0,1}) + (1-\varphi) p(z|t, p_{0,2}), \tag{3.14}$$

i.e. satisfies the superposition principle.

We require $0 \leq \varphi \leq 1$ to ensure that the resulting initial distribution is a probability density, but the relation holds for any φ. The superposition principle will be employed extensively, as it holds for any system properties of interest.

Beyond the probability density of the augmented state, Z_t, the solution of (3.12) might be used to determine the probability density of actual state, X_t. This is achieved be marginalization over θ,

$$p(x|t, p_0) = \int_{\mathbb{R}^{d_Z}} p([x^{\mathrm{T}}, \theta^{\mathrm{T}}]^{\mathrm{T}}|t, p_0) d\theta. \tag{3.15}$$

In much the same way, also the probability density of the measured output, $p(y|t, p_0)$, is obtained. The essential difference is that we do not marginalize individual dimensions. Instead, we marginalize over manifolds, $\{z \in \mathbb{R}_+^{d_Z} | \tilde{\gamma}(z,t) = y\}$, which are defined via a common output, y.

Proposition 3.4. *Given the probability density of the augmented state, $p(z|t, p_0)$, the probability density of the output is:*

$$p(y|t, p_0) = \int_{\{z \in \mathbb{R}_+^{d_Z} | \tilde{y}(z,t)=y\}} p(z|t, p_0) dz. \tag{3.16}$$

As integration is a linear operation, the superposition principle does not only hold for $p(z|t, p_0)$, but also for $p(x|t, p_0)$ and $p(y|t, p_0)$. This is true independent of the integration set, which could be a nonlinear manifold.

3.2.2 Numerical calculation of the population response

To study the dynamics of the cell population, the solution of augmented Fokker-Planck equation has to be computed for a given $p_0(z_0)$. In principle, this can be achieved using common numerical schemes (Patankar, 1980; Strang & Fix, 1973), like finite difference and finite volume methods. Unfortunately, due to the high-dimensionality of (3.12) ($d_Z = d_X + d_\theta \gg 1$) these approaches are prohibitive, as they require a discretization of the space. This is computationally tractable merely for $d_Z \leq 3$.

One alternative to classical grid-based solvers is the method of characteristics (Evans, 1998). This method can be applied if the individual cells have deterministic dynamics, $\tilde{\sigma}(z, t) \equiv 0$ (Weiße *et al.*, 2010). In this case, the probability density of the augmented state, $p(z|t, p_0)$, can be evaluated point-wise by simulation of an ordinary differential equation model. While this is beneficial, the calculation of the output density, $p(y|t, p_0)$ – the output density will later be most important –, requires high-dimensional integration and therefore, many grid points. Also, no extension of this approach to stochastic dynamics is known.

In this thesis, we use a stochastic method to compute $p(z|t, p_0)$ and $p(y|t, p_0)$. This method has been introduced in (Hasenauer *et al.*, 2010b), and employs the fact that the quantities of interest, $p(z|t, p_0)$ and $p(y|t, p_0)$, are probability distributions. While, a direct computation of $p(z|t, p_0)$ and $p(y|t, p_0)$ is inefficient, we can easily sample from these distributions. Therefore, we merely draw a parameter vector and an initial condition from the joint probability density, $p_0(z_0)$, simulate the augmented single cell model for the obtained initial condition, yielding Z_t, and apply the output map, $Y_t = \tilde{y}(Z_t, t)$. If this is done repeatedly, we obtain a sample $\mathcal{S} = \{Y_t^k\}_{k=1}^{d_S}$ from the output density. This sample can be used to approximate output density, $p(y|t, p_0)$, without any evaluation of $p(y|t, p_0)$ (see illustration in Figure 3.3). Therefore, naive density estimators, such as histograms, might be used (as in (Waldherr *et al.*, 2009)), but kernel density estimators outperform those in general.

Kernel density estimators are non-parametric approaches to estimate probability densities from sampled data (Silverman, 1986; Stone, 1984). They are widely used and can be thought of as placing probability "bumps" at each observation, as illustrated in Figure 3.4. These bumps are the kernel functions $\mathbb{K}(y|Y_t^k, H)$, with $\int_{\mathbb{R}^{d_Y}} \mathbb{K}(y|Y_t^k, H)dy = 1$, and added to obtain an estimate,

$$\hat{p}(y|t, \theta) = \frac{1}{d_S} \sum_{i=1}^{d_S} \mathbb{K}(y|Y_t^k, H), \tag{3.17}$$

of $p(y|t, \theta)$. Depending on the problem, different kernel functions might be used. Particularly common are triangular, Epanechnikov, and Gaussian (also called normal) kernels (Silverman,

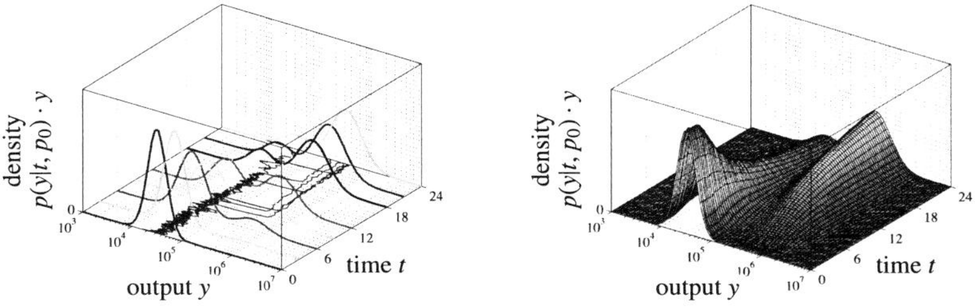

(a) Sampled single cell trajectory (—) and density reconstruction at individual time points (colored lines).

(b) Probability density of output, $p(y|t, p_O)$, describing the output distribution in the cell population.

Figure 3.3: Illustration of single cell trajectories and population density. The population density basically describes the density of single cell trajectories. It can be approximated from a given sample of single cell trajectories using kernel density estimation. (a) and (b) depict simulation results for a bistable stochastic model of the pro-apoptotic signaling pathway (Hasenauer *et al.*, 2011c). Note that the density, $p(y|t, p_O)$, is multiplied by y to compensate for the logarithmic visualization of the output, y.

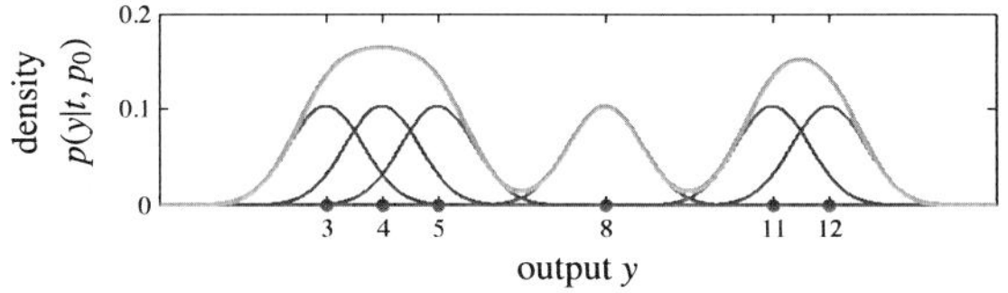

Figure 3.4: Gaussian kernel density estimate (—) of $p(y|t, p_O)$ for the sampled outputs $\bar{Y}_t^k$ (•) and the associated Gaussian kernels (—).

1986). To preserve the positivity of y, multivariate log-normal distributions are used in this thesis,

$$\mathbb{K}(y|Y_t^k, H) = \begin{cases} \dfrac{\exp\left(-\frac{1}{2}(\log y - \log Y_t^k)^T H^{-1} (\log y - \log Y_t^k)\right)}{(2\pi)^{d_Y/2}|H|^{1/2}\prod_{j=1}^{d_Y} y_j} & \text{for } y > 0, \\ 0 & \text{otherwise.} \end{cases} \tag{3.18}$$

The sole parameter of these kernel functions is the positive definite kernel bandwidth matrix $H \in \mathbb{R}^{d_Y \times d_Y}$, also called smoothing parameter (Silverman, 1986; Turlach, 1993). The selection of the smoothing parameter H is crucial and its optimum depends strongly on the size of the sample, d_S, and the distribution of the sample, $\mathcal{S} = \{Y_t^k\}_{i=1}^{d_S}$. A variety of methods exist to select H, ranging from rules-of-thumb – well-known are those by Silverman (1986) and Scott (1992) –, over cross-validation approaches (Stone, 1984), to adaptive bandwidth selection methods (Botev *et al.*, 2010). While rules-of-thumb can be calculated efficiently, cross-validation methods ensure superior asymptotic properties, and adaptive schemes achieve good results for multimodal distributions and in the presence of outliers. In this thesis, we tested different approaches with each example. All depicted results have been obtained using Scott's rule-of-thumb (Scott, 1992),

$$H = \mathrm{diag}(\hat{\sigma}_1^2, \ldots, \hat{\sigma}_{d_Y}^2) d_S^{\frac{1}{d_Y+4}}, \tag{3.19}$$

in which $\hat{\sigma}_j^2$ is the variance estimate for $\log(Y_{j,t})$ determined from $\mathcal{S} = \{Y_t^k\}_{i=1}^{d_S}$. We take here the logarithm as log-normal kernels are used, instead of normal kernels. With this simple formula for H, we achieved reliable results for the considered examples. For other problems more elaborated schemes may be necessary.

By combining the direct sampling with kernel density estimation, we obtain an efficient algorithm for the approximation of $p(y|t, p_0)$. Employing the asymptotic properties of kernel density estimators (see (Silverman, 1986; Stone, 1984)) it can be shown that the expected value of the estimate, $\mathbb{E}[\hat{p}(y|t, p_0)]$, converges to $p(y|t, p_0)$ as d_S increases,

$$\lim_{d_S \to \infty} \mathbb{E}[\hat{p}(y|t, p_0)] = p(y|t, p_0), \tag{3.20}$$

if the kernel bandwidth is adopted appropriately. Via bootstrapping (DiCiccio & Efron, 1996), we can furthermore verify that $\hat{p}(y|t, p_0)$ is close to $p(y|t, p_0)$, which is why this numerical error is disregarded in the following.

Compared to approaches based upon the numerical integration of the partial differential equation and the method of characteristics, our scheme computes the output density, $p(y|t, p_0)$, without calculating the high-dimensional state density, $p(z|t, p_0)$. This is possible as we directly generate a sample from $p(y|t, p_0)$ to which kernel density estimation can be applied. Thus, we circumvent the curse of dimensionality. This is particularly beneficial as the sample size, d_S, required to achieve a good approximation of the state and output density increases with the dimensionality of the corresponding probability distributions (Scott, 1992; Silverman, 1986). For one and two dimensional outputs – the most common number of measurands d_Y – a few thousand trajectories allow in general a good approximation of the output density, $p(y|t, p_0)$ (assuming that no rare events are studied). While this low-dimensionality of the output renders the simulation feasible, still several thousand runs of the single cell model might be required. Therefore, the simulation of the population model is computationally demanding.

3.3 Bayesian estimation and uncertainty analysis of population heterogeneity

In the previous section, an approach to determine the output density within the cell population for a given parameter density, $p_0(z_0)$, was presented. Based upon these results, we approach in this section Problem 3.2, the estimation of $p_0(z_0)$ from the available single cell snapshot data, $\mathcal{D}$.

3.3.1 Bayes' theorem and likelihood function

In the following, we study Problem 3.2 using a Bayesian framework. Therefore, let us recall Bayes' theorem,

$$\pi(p_0|\mathcal{D}) = \frac{\mathbb{P}(\mathcal{D}|p_0)\pi(p_0)}{\mathbb{P}(\mathcal{D})}, \tag{3.21}$$

which states that the posterior probability of $p_0(z_0)$ given $\mathcal{D}$, $\pi(p_0|\mathcal{D})$, is the product of the prior probability of $p_0(z_0)$, $\pi(p_0)$, and the likelihood (or conditional probability) of observing $\mathcal{D}$ given $p_0(z_0)$, $\mathbb{P}(\mathcal{D}|p_0)$, divided by the marginal probability of $\mathcal{D}$, $\mathbb{P}(\mathcal{D}) = \int \mathbb{P}(\mathcal{D}|p_0)\pi(p_0)dp_0$. As discussed in Section 2.3.2 and 2.3.3, the marginal probability of the data is not required for parameter estimation and uncertainty analysis, but we exploit that $\mathbb{P}(\mathcal{D})$ is constant and simply scales $\pi(p_0|\mathcal{D})$, yielding

$$\pi(p_0|\mathcal{D}) \propto \mathbb{P}(\mathcal{D}|p_0)\pi(p_0). \tag{3.22}$$

As the different single cell measurements are independent, the likelihood function (3.22) can be factorized,

$$\pi(p_0|\mathcal{D}) \propto \prod_{i=1}^{d_{\mathcal{D}}} \mathbb{P}(\mathcal{D}^i|p_0)\pi(p_0), \tag{3.23}$$

in which $\mathbb{P}(\mathcal{D}^i|p_0)$ is the likelihood of observing $\mathcal{D}^i$ given $p_0(z_0)$. Note that due to the independence of $\mathcal{D}^{i_1}$ and $\mathcal{D}^{i_2}$, for $i_1 \neq i_2$, it is not necessary to distinguish between the cases that (1) the two cells are measured at the same instance ($t^{i_1} = t^{i_2}$) and that (2) the two cells are measured at different time instances ($t^{i_1} \neq t^{i_2}$). Hence, merely the conditional probability of each individual single cell measurement has to be determined.

For single cell snapshot data, the likelihood of the individual data points, $\mathbb{P}(\mathcal{D}^i|p_0)$, is directly related to $\mathbb{P}(\mathcal{D}^i|X_0, \theta)$, the likelihood of observing $\mathcal{D}^i$ given a single cell with initial state X_0 and parameters θ. This likelihood is

$$\mathbb{P}(\mathcal{D}^i|X_0, \theta) = \int_{\mathbb{R}_+^{d_Y}} p(\bar{Y}^i_{t^i}|y)p(y|t^i, X_0, \theta)dy, \tag{3.24}$$

the marginalization of the probability density of observing $\bar{Y}^i_{t^i}$ given that the single cell has output y, $p(\bar{Y}^i_{t^i}|y)$ – this relates to the measurement noise on the single cell level –, times the probability density $p(y|t^i, X_0, \theta)$ that a single cell with initial state X_0 and parameter θ has output y at time t^i. The probability density of y at time t^i is defined by (2.15), the Fokker-Planck equation for the single cell. Employing the augmented single cell model, we can also write this

$$\mathbb{P}(\mathcal{D}^i|Z_0) = \int_{\mathbb{R}_+^{d_Y}} p(\bar{Y}^i_{t^i}|y)p(y|t^i, Z_0)dy, \tag{3.25}$$

with $\mathbb{P}(\mathcal{D}^i|X_0, \theta) = \mathbb{P}(\mathcal{D}^i|Z_0)$, for $Z_0 = [X_0^T, \theta^T]^T$.

To determine $\mathbb{P}(\mathcal{D}^i|p_0)$ from $\mathbb{P}(\mathcal{D}^i|X_0, \theta)$, we have to weight $\mathbb{P}(\mathcal{D}^i|X_0, \theta)$ with the joint probability density of the initial condition X_0 and the parameters θ, $p_0([x_0^T, \theta^T]^T)$, and marginalize. This yields

$$\mathbb{P}(\mathcal{D}^i|p_0) = \int_{\mathbb{R}_+^{d_X}} \int_{\mathbb{R}_+^{d_\theta}} \mathbb{P}(\mathcal{D}^i|x_0, \theta)p_0([x_0^T, \theta^T]^T)d\theta dx_0, \tag{3.26}$$

$$= \int_{\mathbb{R}_+^{d_X}} \int_{\mathbb{R}_+^{d_\theta}} \int_{\mathbb{R}_+^{d_Y}} p(\bar{Y}^i_{t^i}|y)p(y|t^i, x_0, \theta)dy p_0([x_0^T, \theta^T]^T)d\theta dx_0, \tag{3.27}$$

formulated using X_0 and θ, or

$$\mathbb{P}(\mathcal{D}^i|p_0) = \int_{\mathbb{R}_+^{d_Z}} \int_{\mathbb{R}_+^{d_Y}} p(\bar{Y}^i_{t^i}|y)p(y|t^i, z_0)dy p_0(z_0)dz_0, \tag{3.28}$$

when formulated using the augmented state variable, z_0. Apparently, this integral cannot be evaluated analytically in general. Also numerical methods based on gridding, such as Riemann-Darboux's integration and Lebesgue integration, are not feasible for the high-dimensional problem at hand. Instead, we employ Monte Carlo integration (MacKay, 2005).

Proposition 3.5. *Let $\mathcal{S} = \{Z_0^k\}_{k=1}^{d_S}$ be a representative sample from $p_0(z_0)$, and $\{Y_{t^i}^k\}_{k=1}^{d_S}$ be a set of corresponding stochastic realization of the single cell model (3.9) at time t^i. Then, an approximation of $\mathbb{P}(\mathcal{D}^i|p_0)$ is*

$$\hat{\mathbb{P}}(\mathcal{D}^i|p_0) = \frac{1}{d_S} \sum_{k=1}^{d_S} p(\bar{Y}_{t^i}^i|Y_{t^i}^k), \tag{3.29}$$

with $\lim_{d_S \to \infty} \mathbb{E}\left[\hat{\mathbb{P}}(\mathcal{D}^i|p_0)\right] = \mathbb{P}(\mathcal{D}^i|p_0)$.

Proof. To prove Proposition 3.5 note that (3.28) can be written as

$$\mathbb{P}(\mathcal{D}^i|p_0) = \int_{\mathbb{R}_+^{d_Y}} p(\bar{Y}_{t^i}^i|y) \underbrace{\int_{\mathbb{R}_+^{d_Z}} p(y|t^i, z_0) p_0(z_0) dz_0 \, dy}_{(*)}. \tag{3.30}$$

The inner part, $(*)$, is the probability density that a cell with initial condition z_0 has output y at time t^i, $p(y|t^i, z_0)$, times the probability density of the initial condition z_0, $p_0(z_0)$, integrated with respect to z_0. Hence, $(*)$ is simply the output density $p(y|t^i, p_0)$ of the augmented Fokker-Planck equation, as defined in (3.16), yielding

$$\mathbb{P}(\mathcal{D}^i|p_0) = \int_{\mathbb{R}_+^{d_Y}} p(\bar{Y}_{t^i}^i|y) p(y|t^i, p_0) dy. \tag{3.31}$$

As $\{Y_{t^i}^k\}_{k=1}^{d_S}$ is a sample from $p(y|t^i, p_0)$ (see discussion in Section 3.2.2), we perform here a common Monte Carlo integration (MacKay, 2005). Therefore, this approach inherits all the properties of Monte Carlo integration, among others convergence of the expected value. This concludes the proof. □

Remark 3.6. *We note that in case of small measurement noise levels the Monte-Carlo integral might converge slowly because $p(\bar{Y}_{t^i}^i|y)$ is a narrow distribution and only few samples well contribute the sum in (3.29). In this case it might be computationally more efficient to approximate $\mathbb{P}(\mathcal{D}^i|p_0)$ using a kernel density estimate of $p(\bar{y}|t^i, p^0)$. The methods to derive these kernel density estimate have been introduced in Section 3.2.2 and are discussed in (Hasenauer* et al., *2010b).*

Proposition 3.5 enables the development of a simple algorithm to approximate the likelihood $\mathbb{P}(\mathcal{D}^i|p_0)$ and hence the posterior probability, $\pi(p_0|\mathcal{D})$. The pseudocode is provided in Algorithm 3.1. Note that the sample $\{p(\bar{Y}_{t^i}^i|Y_{t^i}^k)\}_{k=1}^{d_S}$ generated for the calculation of $\hat{\mathbb{P}}(\mathcal{D}^i|p_0)$ can also be used to determine the uncertainty of $\hat{\mathbb{P}}(\mathcal{D}^i|p_0)$ (MacKay, 2005). In the following, we ensure that this uncertainty is small, meaning that the approximation error can be neglected.

While proving Proposition 3.5, we also established a link between the population density and the noise corrupted single cell measurement. Eq. (3.31) verifies that the likelihood can be defined merely using population level information. This is crucial, as a single cell description

Algorithm 3.1 Pseudocode for Monte Carlo integration of likelihood.

Require: Single cell model, data point $\mathcal{D}^i = (\bar{Y}^i_{t^i}, t^i)$, initial density $p_0(z_0)$, sample size d_S.
  for $k = 1$ to d_S **do**
    Draw Z^k_0 from $p_0(z_0)$, $Z^k_0 \sim p_0(z_0)$.
    Compute realization, $Z^k_{t^i}$, of single cell model for initial condition Z^k_0.
    Evaluate likelihood $p^i_k = p(\bar{Y}^i_{t^i}|\tilde{y}(Z^k_{t^i}))$.
  end for
  Compute mean of likelihoods: $\hat{\mathbb{P}}(\mathcal{D}^i|p_0) = \frac{1}{d_S}\sum_{k=1}^{d_S} p^i_k$, $i = 1, \ldots, d_{\mathcal{D}}$.

of the process is not feasible in case of single cell snapshot data. This is in contrast to single cell time-lapse data for which the number of measured cells is a lot smaller, but much more information is available per individual cell. For single cell snapshot data, a parameterization of the individuals, as in (Koeppl *et al.*, 2012), would result in thousands of unknown parameters and extreme identifiability problems. The parameter vectors of the individual cells would only be restricted by a single measurement. The exploration of the resulting posterior would be challenging also for advanced sampling schemes (as introduced in Section 2.3.3).

Based on (3.21) & (3.29), the calculation of the posterior probability for a given probability density of the parameters is possible. Unfortunately, the computational effort associated to the evaluation of the likelihood function might be large, as Algorithm 3.1 requires many simulations of the single cell model. This renders the repeated evaluation of $\mathbb{P}(\mathcal{D}^i|p_0)$ prohibitive. The problem is complicated further as $p_0(z_0)$ is a function. The corresponding inverse problem (Problem 3.2) of estimating $p_0(z_0)$ from $\mathcal{D}$ is therefore infinite dimensional.

3.3.2 Parametric distribution for parameters and initial conditions

To elude the infinite dimensional inference problem, the parameter density is parameterized.

Definition 3.7. *Let* $\forall j \in \{1, \ldots, d_\varphi\} : \Lambda_j : \mathbb{R}^{d_z}_+ \to \mathbb{R}_+$ *be a multivariate probability density,* $\int_{\mathbb{R}^{d_z}_+} \Lambda_j(z_0)dz_0 = 1$. *Then, for* $\varphi \in [0, 1]^{d_\varphi}$ *and* $\mathbf{1}^{\mathrm{T}}\varphi = 1$,

$$p_{0,\varphi}(z_0) = \sum_{j=1}^{d_\varphi} \varphi_j \Lambda_j(z_0), \tag{3.32}$$

is called a linear parametric model of $p_0(z_0)$.

The multivariate probability densities, Λ_j, are denoted as ansatz functions. Possible choices are, for instance, Gaussian distributions and head-type functions. The theoretical results presented in the following are independent of the choice of the ansatz functions, as long as they are probability densities. The weightings φ_j of the ansatz functions can be interpreted as parameters determining the probability density $p_{0,\varphi}(z_0)$ and are denoted as density parameters. The density parameters are the unknown variables when inferring the deterministic source of population heterogeneity. We will only allow for consistent density parameters:

Definition 3.8. *A parameter vector* φ *is called consistent density parameter vector if:*

1. *all entries of* φ *are non-negative,* $\varphi \in [0, 1]^{d_\varphi}$, *and*
2. *the sum of all entries is one,* $\mathbf{1}^{\mathrm{T}}\varphi = 1$.

For consistent density parameters the corresponding distribution $p_{0,\varphi}(z_0)$ is a probability density, $\forall z_0 \in \mathbb{R}_+^{d_Z} : p_{0,\varphi}(z_0) \geq 0$ and $\int_{\mathbb{R}_+^{d_Z}} p_{0,\varphi}(z_0) dz_0 = 1$.

Given a parameterization $p_{0,\varphi}$, the output density can be simplified according to the following corollary to Theorem 3.2 and Proposition 3.3.

Corollary 3.9. *Let* $\forall j \in \{1, \ldots, d_\varphi\} : p(z|t, \Lambda_j)$ *be the solution of the augmented Fokker-Planck equation* (3.12) *and* $p(y|t, \Lambda_j)$ *the output* (3.4) *for initial probability density* $\Lambda_j(z_0)$. *Then, the solution of* (3.12) *&* (3.4) *it holds:*

$$p(z|t, p_{0,\varphi}) = p\left(z|t, \sum_{j=1}^{d_\varphi} \varphi_j \Lambda_j\right) = \sum_{j=1}^{d_\varphi} \varphi_j p(z|t, \Lambda_j). \tag{3.33}$$

and

$$p(y|t, p_{0,\varphi}) = p\left(y|t, \sum_{j=1}^{d_\varphi} \varphi_j \Lambda_j\right) = \sum_{j=1}^{d_\varphi} \varphi_j p(y|t, \Lambda_j). \tag{3.34}$$

The proof of this proposition is straight forward. It employs the superposition principle (Proposition 3.3) and the linearity of the population level output map (3.4). The parametric formulation is advantageous as the computation of the output density for arbitrary density parameters, φ, only requires the non-recurring computation of the responses $p(y|t, \Lambda_j)$ and the weighted summation of those.

Note that biologically, the different ansatz functions might be interpreted as properties of subpopulations. In subpopulation j, parameters and initial conditions are distributed according to $\Lambda_j(z_0)$. In this context, the density parameter φ_j is the contribution of the j-th subpopulation to the complete population. To analyze the population heterogeneity, its subpopulation composition, φ, must be assessed.

Generalized gray-box parameterization

The parameterization presented above can be generalized in various ways. Of particular practical relevance are situations in which the distribution of some states or parameters is known. This knowledge should certainly be used to reduce the problem's complexity. This can be achieved by particular selection approaches for ansatz functions.

For illustration purposes we assume that the l-th component of Z_0, $Z_{l,0}$, is independent of the other variables and distributed according to $p_{l,0}(z_{l,0})$,

$$\begin{aligned} p_0([z_{1,0}, \ldots, z_{l-1,0}, z_{l,0}, z_{l+1,0}, \ldots, z_{d_Z,0}]) = \\ p_{\{1,\ldots,l-1,l+1,\ldots d_Z\},0}([z_{1,0}, \ldots, z_{l-1,0}, z_{l+1,0}, \ldots, z_{d_Z,0}]) p_{l,0}(z_{l,0}). \end{aligned} \tag{3.35}$$

This information can be employed to reduce the dimensionality of the ansatz functions. Instead of parameterizing the whole distribution, only the unknown components are parameterized,

$$p_{0,\varphi}(z_0) = \sum_{j=1}^{d_\varphi} \varphi_j \Lambda_j(z_0) = \left(\sum_{j=1}^{d_\varphi} \varphi_j \tilde{\Lambda}_j([z_{1,0}, \ldots, z_{l-1,0}, z_{l+1,0}, \ldots, z_{d_Z,0}]^T)\right) p_{l,0}(z_{l,0}). \tag{3.36}$$

This ensures that (3.35) holds, but allows for flexibility in the unknown directions. Such a reduced parameterization still results in a linear parametric model according to Definition 3.8. Similar approaches can be employed if there exist known co-dependency among several components of z_0.

Remark 3.10. *In this thesis we choose a parametric approach and thereby circumvent optimization over a function space. Complementary methods could be based on variational calculus (Fox, 1987). While interesting, this was beyond the scope of this work.*

3.3.3 Exact parameterization of likelihood and posterior probability

In the last section, a parameterization of $p_{0,\varphi}(z_0)$ in terms of φ is introduced. Beyond the reduction of the problem dimension, the linear parametric form of $p_{0,\varphi}(z_0)$ allows for the parametrization of the likelihood function $\mathbb{P}(\mathcal{D}|p_{0,\varphi})$.

Theorem 3.11. *Let $\mathcal{D}$ be a collection of single cell snapshot data, $p_{0,\varphi}$ a linear parametric model of the joint distribution of parameters and initial states, and φ a consistent density parameter. Then, the likelihood of observing $\mathcal{D}$ given $p_{0,\varphi}$ is*

$$\mathbb{P}(\mathcal{D}|p_{0,\varphi}) = \prod_{i=1}^{d_{\mathcal{D}}} \varphi^{\mathrm{T}} c^i, \tag{3.37}$$

in which $c^i = [c_1^i, \dots, c_{d_\varphi}^i]^{\mathrm{T}}$. The elements c_j^i of c^i denote the conditional probability density (likelihood) of observing data point $\mathcal{D}^i$ for a population with initial condition $\Lambda_j(z_0)$,

$$\forall i \in \{1, \dots, d_{\mathcal{D}}\}, j \in \{1, \dots, d_\varphi\}: \quad c_j^i = \mathbb{P}(\mathcal{D}^i|\Lambda_j) = \int_{\mathbb{R}_+^{d_Y}} p(\bar{Y}_{t^i}^i|y)p(y|t, p_{0,\varphi})dy. \tag{3.38}$$

Proof. To prove Theorem 3.11, note that the parameterization of the output density enables the reformulation of the likelihood (3.31),

$$\mathbb{P}(\mathcal{D}^i|p_{0,\varphi}) = \int_{\mathbb{R}_+^{d_Y}} p(\bar{Y}_{t^i}^i|y)p(y|t, p_{0,\varphi})dy \tag{3.39}$$

$$= \int_{\mathbb{R}_+^{d_Y}} p(\bar{Y}_{t^i}^i|y)\left(\sum_{j=1}^{d_\varphi} \varphi_j p(y|t, \Lambda_j)\right)dy. \tag{3.40}$$

The latter allows for the interchange of summation and integration,

$$\mathbb{P}(\mathcal{D}^i|p_{0,\varphi}) = \sum_{j=1}^{d_\varphi} \varphi_j \underbrace{\int_{\mathbb{R}_+^{d_Y}} p(\bar{Y}_{t^i}^i|y)p(y|t, \Lambda_j)dy}_{=:c_j^i}. \tag{3.41}$$

Employing the definition (3.38) of c_j^i and $\mathbb{P}(\mathcal{D}|p_{0,\varphi}) = \prod_{i=1}^{d_{\mathcal{D}}} \mathbb{P}(\mathcal{D}^i|p_{0,\varphi})$ we obtain (3.37), which concludes the proof. □

Formulation (3.37) is superior to formulations (3.28) & (3.31). While changes in the distribution p_0 require the complete reevaluation of (3.28) & (3.31) – this includes sampling and simulation which is computationally demanding –, this is not the case for (3.37). Employing Theorem 3.11, the evaluation of the likelihood function becomes a two-step procedure. In the first step, the so called pre-estimation step, the conditional probabilities, c_j^i, are evaluated. Therefore, Monte Carlo integration is performed using Algorithm 3.1 with density $\Lambda_j(z_0)$. As the ansatz functions are predefined and independent of the unknowns, φ, we calculate c_j^i just once. The second step takes place during the optimization or uncertainty analysis. In this step, the precomputed c_j^i are used to determine the likelihood for a certain φ. This evaluation is extremely fast, as (3.37) provides a simple analytical expression. This analytical expression provides the exact value of the likelihood presuming the likelihoods c_j^i are correct.

These advantages extend to the posterior probability, $\pi(p_{0,\varphi}|\mathcal{D})$. Employing (3.37), we can rewrite the unnormalized posterior:

Corollary 3.12. *Let the conditions of Theorem 3.11 hold and $\pi(p_0)$ be the prior probability. Then, the posterior probability is*

$$\pi(p_{0,\varphi}|\mathcal{D}) \propto \left(\prod_{i=1}^{d_{\mathcal{D}}} \varphi^{\mathrm{T}} c^i\right) \pi(\varphi), \tag{3.42}$$

with c_j^i defined according to (3.38) *and* $\pi(\varphi) := \pi(p_{0,\varphi})$.

This follows directly from Theorem 3.11, and the definition of the unnormalized posterior (3.22).

3.3.4 Computation of maximum a posteriori estimate

Given the likelihood function, an essential question is which density $p_{0,\varphi^*}(z_0)$ maximizes the posterior probability.

Definition 3.13. *Let the conditions of Corollary 3.12 be fulfilled. Then, the consistent density parameter vector, φ^*, of the maximum a posteriori estimate, $p_{0,\varphi^*}(z_0)$, is the optimal solution to*

$$\begin{aligned} &\underset{\varphi \in [0,1]^{d_\varphi}}{\text{maximize}} && \left(\prod_{i=1}^{d_{\mathcal{D}}} \varphi^{\mathrm{T}} c^i\right) \pi(\varphi) \\ &\text{subject to} && \mathbf{1}^{\mathrm{T}} \varphi = 1. \end{aligned} \tag{3.43}$$

Here, the constraints ensure that the obtained density is positive and has integral one. Hence, only consistent density parameter vector, φ^*, are feasible. The density $p_{0,\varphi^*}(z_0)$ resulting for φ^* is the most likely joint density of parameters and initial condition given the measured data and the prior knowledge. Employing the parametric form of the posterior probability provided by (3.42), this maximum a posteriori estimate can be computed using global nonlinear optimization (see, e.g., (Weise, 2009) and references therein). This computation is rather efficient as the computational cost of evaluating the objective is negligible. This is due to the parameterization of the likelihood function.

Still, if a large number of ansatz functions is used, the computation time associated to global optimization might be large. Then, a restriction of the prior probability to the quasiconcave functions (Boyd & Vandenberghe, 2004) can be helpful.

Definition 3.14. *A prior probability $\pi(\varphi) : \mathbb{R}^{d_\varphi} \to \mathbb{R}_+$ is called quasiconcave (or unimodal) if its sublevel sets*

$$\Omega_\alpha = \{\varphi \in [0,1]^{d_\varphi} | \pi(\varphi) \geq \alpha\}, \tag{3.44}$$

are convex sets for all $\alpha \in \mathbb{R}_+$.

In general, the restriction to quasiconcave prior probabilities is not critical. Most of the common multivariate priors are quasiconcave, among others, products of unimodal one-dimensional priors and multivariate Gaussian distribution. The confinement to quasiconcave priors offers tremendous advantages concerning the computation time.

Theorem 3.15. *Let the prior $\pi(\varphi)$ be quasiconcave and c^i be defined by* (3.38). *Then, the maximum a posteriori estimate, φ^*, is the solution to the quasiconvex optimization problem*

$$\begin{aligned} &\underset{\varphi \in [0,1]^{d_\varphi}}{\text{minimize}} \quad -\left[\sum_{i=1}^{d_\mathcal{D}} \log(\varphi^\mathrm{T} c^i) + \log(\pi(\varphi))\right] \\ &\text{subject to} \quad \mathbf{1}^\mathrm{T}\varphi = 1. \end{aligned} \tag{3.45}$$

Proof. To prove Theorem 3.15, we transform the problem by applying the negative logarithm to the objective function. The maximization becomes a minimization and we obtain (3.45). The transformed problem has a convex domain, $\varphi \in [0,1]^{d_\varphi}$, and a quasiconvex objective function. To verify the latter, note that the application of the logarithm preserves quasiconcavity, as the sublevel sets remain unchanged. Therefore, $\log(\pi(\varphi))$ and $\log(\varphi^\mathrm{T} c^i)$ are quasiconcave. As the sum of quasiconcave functions is also quasiconcave, $\sum_{i=1}^{d_\mathcal{D}} \log(\varphi^\mathrm{T} c^i)$ is quasiconcave and therefore the whole expression $\left[\sum_{i=1}^{d_\mathcal{D}} \log(\varphi^\mathrm{T} c^i) + \log(\pi(\varphi))\right]$. Employing that the negation of a quasiconcave function is a quasiconvex function, we arrive at Theorem 3.15, which concludes the proof. □

As $-\left[\sum_{i=1}^{d_\mathcal{D}} \log(\varphi^\mathrm{T} c^i) + \log(\pi(\varphi))\right]$ is quasiconcave, $\left[\sum_{i=1}^{d_\mathcal{D}} \log(\varphi^\mathrm{T} c^i) + \log(\pi(\varphi))\right]$ is quasiconvex (Boyd & Vandenberghe, 2004, Section 3.4.1). Accordingly, also the posterior distribution $\pi(p_{0,\varphi}|\mathcal{D}) \propto \left(\prod_{i=1}^{d_\mathcal{D}} \varphi^\mathrm{T} c^i\right)\pi(\varphi) = \exp\left[\sum_{i=1}^{d_\mathcal{D}} \log(\varphi^\mathrm{T} c^i) + \log(\pi(\varphi))\right]$ is quasiconvex as the exponential is a strictly monotone function which merely shifts the sublevel sets ($\Omega_\alpha \rightarrow \Omega_{\exp(\alpha)}$). For details on concavity, convexity and the conservation of these properties, we refer to Boyd & Vandenberghe (2004). These authors provide an in-depth discussion of convex optimization problems and reformulation approaches to obtain those.

Quasiconvexity of the optimization problem – as established by Theorem (3.15) – is very valuable. It allows us to solve (3.45) using a sequence of convex optimization problems (Boyd & Vandenberghe, 2004; Luenberger, 1968). This sequence is constructed via bisection over the objective function value. In each step of the sequence, a convex optimization problem is solved. For convex optimization problems there exist solvers which ensure convergence to the global optimum in polynomial time, e.g, the interior point methods (Boyd & Vandenberghe, 2004; Löfberg, 2009). As the number of bisections required to determine the optimum to numerical precision is generally small, sequential convex optimization is far more efficient than global nonlinear optimization. In addition, convergence to the global optimum is guaranteed (Boyd & Vandenberghe, 2004).

Remark 3.16. *The proof of quasiconcavity of the objective function of* (3.43)*, respectively the quasiconcavity of $\pi(p_{0,\varphi}|\mathcal{D}) \propto \left(\prod_{i=1}^{d_\mathcal{D}} \varphi^\mathrm{T} c^i\right)\pi(\varphi)$, also establishes unimodality of the posterior distribution $\pi(p_{0,\varphi}|\mathcal{D})$ (Boyd & Vandenberghe, 2004, Section 3.4.1).*

The efficient estimation of the global optimum of $p_{0,\varphi}(z_0)$ also enables the computation of $p_{\theta,\varphi}(\theta)$. Therefore, merely marginalization over x_0 is necessary,

$$p_{\theta,\varphi}(\theta) = \int_{\mathbb{R}_+^{d_x}} p_{0,\varphi}([x_0^\mathrm{T}, \theta^\mathrm{T}]^\mathrm{T}) dx_0. \tag{3.46}$$

3.3.5 Bayesian analysis of the model uncertainties

In the previous section a method was presented which allows the computation of the maximum a posteriori estimate, $p_{0,\varphi^*}(z_0)$. As measurement data are limited and noise corrupted, this estimate will not always reflect the true parameter density. Hence, the uncertainty of the parameter density has to be evaluated.

MCMC sampling of the posterior distribution

To analyze the uncertainty of the density estimate, a sample of the posterior, $\pi(p_{0,\theta}|\mathcal{D})$, is collected. This is possible, as the unnormalized posterior probability of a distribution can be evaluated efficiently given (3.42). In this work, the sampling is performed with a Metropolis-Hastings method (see Section 2.3.3). Also Gibbs or slice sampling approaches may be employed. Compared to importance and rejection sampling these methods are well suited as they do not require the selection of an appropriate global proposal density, a task which is difficult for the considered problem.

Markov chain Monte Carlo (MCMC) sampling merely requires the selection of a local proposal (or transition) density. Still, although the problem at hand possesses a unimodal posterior distribution, MCMC sampling is not straight forwards as we have to ensure consistency of the density parameters. They have to be non-negative and the elements have to sum up to one. Due to the latter, prior probability and posterior probability are non-zero only on a $(d_\varphi - 1)$-dimensional subset of the density parameter space. If for such a problem a standard proposal step is used, the acceptance rate would be zero.

We overcome this challenge by performing the sampling in the $(d_\varphi - 1)$-dimensional space, $[\varphi_1, \dots, \varphi_{d_\varphi - 1}]^{\mathrm{T}} \in \mathbb{R}^{d_\varphi - 1}$. The remaining density parameter is determined via the closing condition $\varphi_{d_\varphi} = 1 - \sum_{j=1}^{d_\varphi} \varphi_j$. According, the update step consists of two sub-steps:

1. Draw proposals $[\varphi_1^{k+1}, \dots, \varphi_{d_\varphi-1}^{k+1}]^{\mathrm{T}}$ from the $(d_\varphi - 1)$–dimensional reduced proposal density $\mathbb{Q}_r$,

$$\begin{bmatrix} \varphi_1^{k+1} \\ \vdots \\ \varphi_{d_\varphi-1}^{k+1} \end{bmatrix} \sim \mathbb{Q}_r\left(\begin{bmatrix} \varphi_1^{k+1} \\ \vdots \\ \varphi_{d_\varphi-1}^{k+1} \end{bmatrix} \middle\| \begin{bmatrix} \varphi_1^{k} \\ \vdots \\ \varphi_{d_\varphi-1}^{k} \end{bmatrix}\right). \tag{3.47}$$

2. Determine $\varphi_{d_\varphi}^{k+1}$ such that $\sum_{j=1}^{d_\varphi} \varphi_j^{k+1} = 1$,

$$\varphi_{d_\varphi}^{k+1} = 1 - \sum_{j=1}^{d_\varphi - 1} \varphi_j^{k+1}. \tag{3.48}$$

This two-step procedure can also be written as a single sampling step, $\varphi^{k+1} \sim \mathbb{Q}(\varphi^{k+1}, \varphi^k)$, for an appropriate choice of $\mathbb{Q}$. As usual, after the parameter vector φ^{k+1} has been proposed it is accepted or rejected. The accepted probability is

$$p_a = \min\left\{1, \frac{\pi(p_{0,\varphi^{k+1}}|\mathcal{D})}{\pi(p_{0,\varphi^k}|\mathcal{D})} \frac{\mathbb{Q}(\varphi^k, \varphi^{k+1})}{\mathbb{Q}(\varphi^{k+1}, \varphi^k)}\right\}, \tag{3.49}$$

$$= \min\left\{1, \frac{\prod_{i=1}^{d_\mathcal{D}} (\varphi^{k+1})^{\mathrm{T}} c^i}{\prod_{i=1}^{d_\mathcal{D}} (\varphi^{k})^{\mathrm{T}} c^i} \frac{\pi(\varphi^{k+1})}{\pi(\varphi^k)} \frac{\mathbb{Q}(\varphi^k, \varphi^{k+1})}{\mathbb{Q}(\varphi^{k+1}, \varphi^k)}\right\}. \tag{3.50}$$

This acceptance probability has the common structure but makes use of the parametric likelihood. In order to rule out $\varphi_{d_\varphi}^{k+1} < 0$ we require that the prior probability $\pi(\varphi^{k+1})$ is zero if φ^{k+1} is not a consistent density parameter.

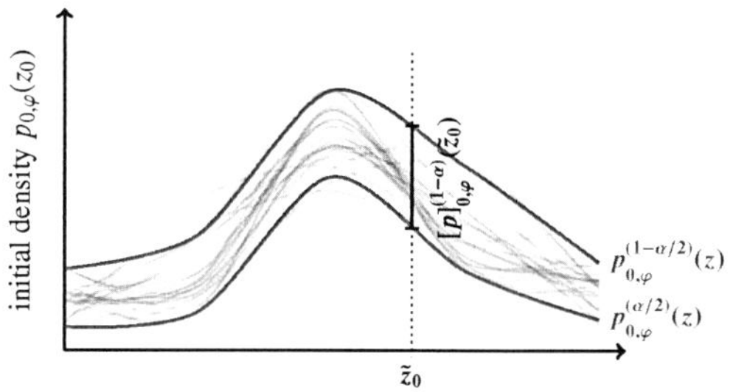

Figure 3.5: Illustration of a function sample $\{p_{0,\varphi^k}(z_0)\}_{k=1}^{d_S}$ ( ), percentile functions $p^{(\alpha)}_{0,\varphi}(z_0)$ (—), and the Bayesian confidence interval at a particular point $\tilde{z}_0$, $[p]^{(1-\alpha)}_{0,\varphi}(\tilde{z}_0)$ (—).

Combining update and acceptance step, we obtain a MCMC algorithm with the same structure as Algorithm 2.1. This algorithm draws a sample of consistent density parameters, $\{\varphi^k\}_{k=1}^{d_S}$, or respectively joint densities of parameter and initial state, $\{p_{0,\varphi^k}(z_0)\}_{k=1}^{d_S}$, from the posterior distribution. The facts that

- the conditional probabilities c^i_j are only computed once in the beginning, and that
- every evaluation of the acceptance probability p_a requires only a small number of algebraic operations,

ensure hereby an efficient sampling. Without this reformulation, the integral defining the likelihood $\mathbb{P}(\mathcal{D}|p_{0,\varphi})$ would have to be evaluated in each update step. The resulting computational effort would be restrictive for practical applications.

Bayesian confidence intervals for model parameterization

The sample $\{p_{0,\varphi^k}(z_0)\}_{k=1}^{d_S}$ collected by the MCMC algorithm contains information about the shape of the posterior density. This information can be employed to determine the Bayesian confidence intervals, here defined via percentiles of the sample (see Section 2.3.2).

Commonly, the 100α-th percentile is a single number. For the problem of density estimation it is however a function. As we are interested in the confidence intervals of $p_{0,\varphi}(z_0)$, the percentiles are defined point-wise for every z_0. The 100α-th percentile of $p_{0,\varphi}(z_0)$ is the function $p^{(\alpha)}_{0,\varphi}(z_0)$ below which $100\alpha\%$ of the observations $p_{0,\varphi^k}(z_0)$ fall for each z,

$$\forall z_0 \in \mathbb{R}^{d_z}_+ : \quad \Pr(p_{0,\varphi}(z_0) \leq p^{(\alpha)}_{0,\varphi}(z_0)) = \alpha. \tag{3.51}$$

Note that the function $p^{(\alpha)}_{0,\varphi}(z_0)$ is in general not a probability density, as the integral value is not one.

Based upon (3.51), the $100(1-\alpha)\%$ Bayesian confidence interval $[p]^{(1-\alpha)}_{0,\varphi}(z_0)$ of $p_{0,\varphi}(z_0)$ is defined as

$$[p]^{(1-\alpha)}_{0,\varphi}(z_0) = [p^{(\alpha/2)}_{0,\varphi}(z_0), p^{(1-\alpha/2)}_{0,\varphi}(z_0)]. \tag{3.52}$$

As a sample $\{p_{0,\varphi^k}\}_{k=1}^{d_S}$ is available, an approximation of $p^{(\alpha/2)}_{0,\varphi}(z_0)$ and $p^{(1-\alpha/2)}_{0,\varphi}(z_0)$ can be determined using the percentiles of the sample. Approximation of the percentile values can be computed using, for instance, the nearest rank method or linear interpolation between closest ranks. The relation between sample, percentiles, and confidence intervals is sketched in Figure 3.5.

Similar to $p_{0,\varphi}(z_0)$, Bayesian confidence intervals might be defined and computed for every marginalization of $p_{0,\varphi}(z_0)$. In particular confidence interval of the parameter densities, $p_{\theta,\varphi}(\theta)$,

$$[p]_{\theta,\varphi}^{(1-\alpha)}(\theta) = [p_{\theta,\varphi}^{(\alpha/2)}(\theta), p_{\theta,\varphi}^{(1-\alpha/2)}(\theta)]. \tag{3.53}$$

might be of interest. To compute this confidence interval, at first the marginalization (3.46) of $p_{0,\varphi^k}(z_0)$ over x_0 has to be performed for $k = 1, \ldots, d_S$. This requires high-dimensional integration and might be computationally demanding.

Bayesian confidence intervals for the model prediction

Beyond the analysis of the state and parameter distribution, the response of the biological system under altered conditions and the dynamics of unmeasured (hidden) states is of interest. This response can be predicted using the augmented Fokker-Planck equation. The simplest approach is the simulation of the system using the maximum a posteriori estimate, $p_{0,\varphi^*}(z_0)$. This prediction can be written as

$$p(y|t, p_{0,\varphi^*}) = \sum_{j=1}^{d_\varphi} \varphi_j^* p(y|t, \Lambda_j), \tag{3.54}$$

with $p(y|t, \Lambda_j)$ being the response of a hypothetical population possessing parameter distribution $\Lambda_j(z_0)$.

Unfortunately, this simple approach based upon the maximum a posterior estimate has one severe shortcoming. As the parameter density is not known precisely, also the model predictions are uncertain. To evaluate the reliability of the population model and its predictive power, these prediction uncertainties have to be quantified for the experiment and the property of interest. Therefore, we use time-dependent Bayesian confidence intervals.

The $100(1-\alpha)\%$ confidence intervals of the predictions $[p]_{y,\varphi}^{(1-\alpha)}(y|t)$ are defined via the percentile method,

$$[p]_{y,\varphi}^{(1-\alpha)}(y|t) = [p^{(\alpha/2)}(y|t, p_{0,\varphi}), p^{(1-\alpha/2)}(y|t, p_{0,\varphi})], \tag{3.55}$$

as before. Therein, the 100α-th percentile $p^{(\alpha)}(y|t, p_{0,\varphi})$ of the predicted output $p(y|t, p_{0,\varphi})$ is

$$\forall y \in \mathbb{R}_+^{d_Y},\ \forall t \in \mathbb{R}_+ : \quad \Pr(p(y|t, p_{0,\varphi}) \leq p^{(\alpha)}(y|t, p_{0,\varphi})) = \alpha. \tag{3.56}$$

To compute $p^{(\alpha)}(y|t, p_{0,\varphi})$, the output densities are computed (see Section 3.2.2). These output densities are used together with the sample from the posterior density, $\{p_{0,\varphi^k}(z_0)\}_{k=1}^{d_S}$ of the posterior density, to obtain a sample from the posterior output density, $\pi(p(y|t, p_{0,\varphi})|\mathcal{D})$. This sample is obtained be weighting the prediction for $\Lambda_j(z_0)$ with φ_j,

$$\left\{p(y|t, p_{0,\varphi^k})\right\}_{k=1}^{d_S} = \left\{\sum_{j=1}^{d_\varphi} \varphi_j^k p(y|t, \Lambda_j)\right\}_{k=1}^{d_S}. \tag{3.57}$$

The obtained sample is a representative sample form $\pi(p(y|t, p_{0,\varphi})|\mathcal{D})$ and can be used to approximate the prediction confidence interval $[p]_{y,\varphi}^{(1-\alpha)}(y|t)$. As the population model has to be simulated only d_φ times, this calculation is computationally tractable.

This approach can be easily generalized to state distribution, linear combinations of states and other properties. Therefore, this procedure allows for the direct assessment of uncertainties in population models and supports the selection of future experiments.

To sum up, in this section a method for the estimation of parameter distributions in heterogeneous cell populations from population data has been presented. The approach relies on the linear parameterization of the parameter heterogeneity, which allows for the parametrization of the likelihood function. It has been shown that the optimal value as well as the Bayesian confidence intervals can be computed efficiently employing a parameterization of the parameter density. Also a method to determine prediction uncertainties has been presented. This allows an in-depth analysis of the reliability of the model. A summary of the procedure is shown in Figure 3.6.

3.4 Example: Apoptotic signaling in cancer cell populations

In multicellular organisms, the removal of infected, malfunctioning, or no longer needed cells is important to avoid, e.g., cancer progression (Gewirtz *et al.*, 2007; Spencer & Sorger, 2011). Therefore, multicellular organisms developed different mechanisms to externally enforce cell death. Key players in this process are signaling molecules belonging to the TNF family, such as TNF-α, TNF-β, and TRAIL (Wajant *et al.*, 2003). These molecules bind to specific death receptors in the cell membrane and induce apoptosis – one type of programmed cell death – via the activation of the caspase cascade. On the other hand, they can also promote cell survival via the inflammatory response, specifically activation of the NF-κB pathway (Wajant *et al.*, 2003). The proportion of the activation of these two signaling pathways decides about the fate of the single cell.

The importance of pro- and antiapoptotic signaling let to the development of a variety of models (see, e.g., (Cheong *et al.*, 2008; Kalita *et al.*, 2011; Lipniacki *et al.*, 2004; Paszek *et al.*, 2010; Waldherr, 2009) for antiapoptotic signaling and (Albeck *et al.*, 2008a,b; Eissing *et al.*, 2009, 2004; Spencer *et al.*, 2009; Spencer & Sorger, 2011; Witt *et al.*, 2009) for proapoptotic signaling), most of which consider only one of the signaling pathways. In the following, a simple model for the caspase and NF-κB activation is studied which reproduces the main features of the single cell response to a TNF-α stimulus. To illustrate the properties of the proposed methods, we consider at first a case where only one parameter is distributed. In a second step, we show that the method is also applicable in the case of multi-parametric heterogeneity.

3.4.1 Model of the TNF signaling pathway

The model considered in this study has been introduced in (Chaves *et al.*, 2008) and accounts for known activating and inhibitory interactions among key signaling proteins of the TNF pathway. A schematic is shown in Figure 3.7. Besides active caspase 8 (C8a) and active caspase 3 (C3a), the nuclear transcription factor κB (NF-κB) and its inhibitor I-κB are considered

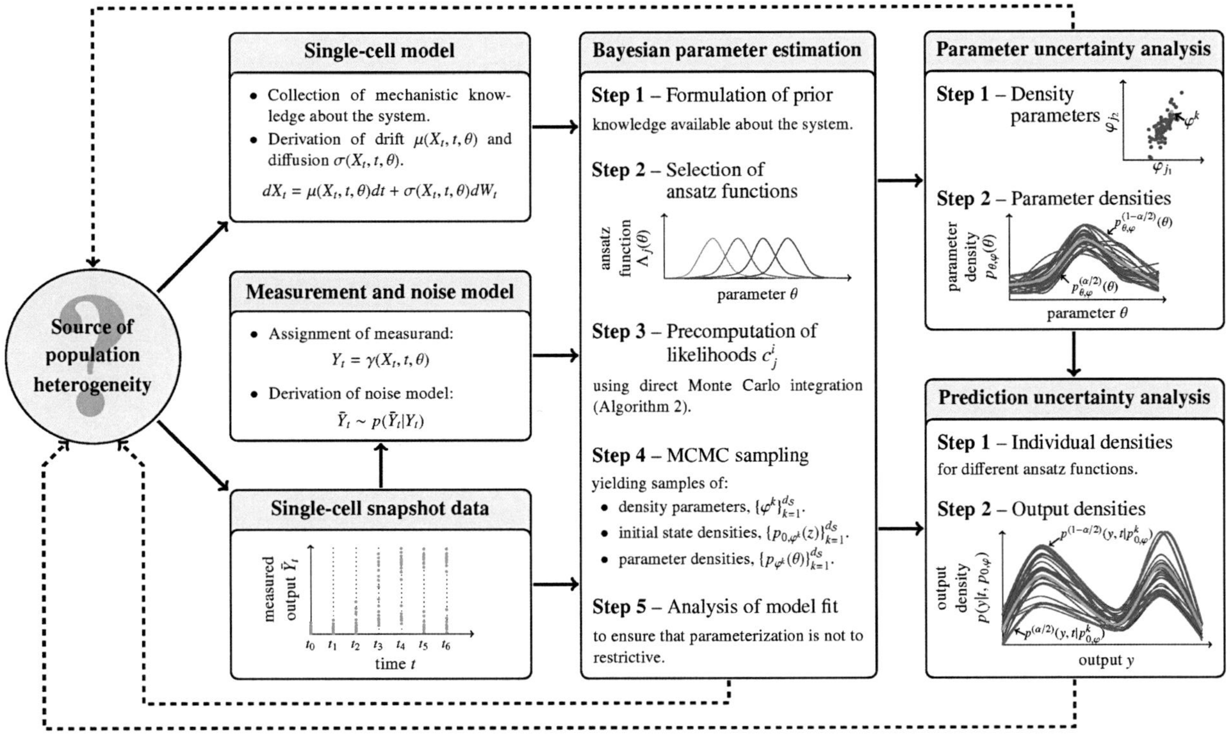

Figure 3.6: Illustration of the proposed Bayesian parameter estimation procedure. The main steps as well as their order is shown. Note in particular that simulations of the single cell model as well as the population model are only required for the pre-computation of the likelihood (Step 3), while the MCMC sampling (Step 4) merely employs the precomputed c^i_j's.

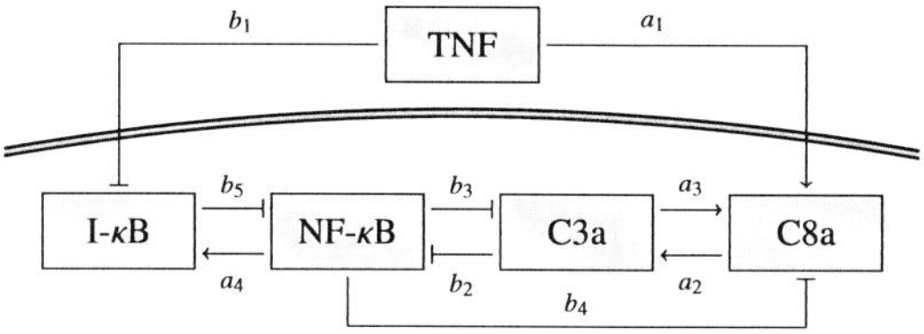

Figure 3.7: Graphical representation of the TNF signal transduction model. Activation is indicated by → and inhibition by ⊣.

Table 3.1: Nominal parameter values for the TNF signaling model (3.58).

i	1	2	3	4	5
a_i	0.6	0.2	0.2	0.5	
b_i	0.4	0.7	0.3	0.5	0.4

in the model. The model is given by the ODE system

$$\begin{aligned} dX_{1,t} &= (-X_{1,t} + 0.5\,(\text{inh}_4(X_{3,t})\text{act}_1(u) + \text{act}_3(X_{2,t})))\,dt \\ dX_{2,t} &= (-X_{2,t} + \text{act}_2(X_{1,t})\text{inh}_3(X_{3,t}))\,dt \\ dX_{3,t} &= (-X_{3,t} + \text{inh}_2(X_{2,t})\text{inh}_5(X_{4,t}))\,dt \\ dX_{4,t} &= (-X_{4,t} + 0.5\,(\text{inh}_1(u) + \text{act}_4(X_{3,t})))\,dt. \end{aligned} \tag{3.58}$$

The state variables $X_{i,t}$, $i = 1, \ldots, 4$, denote the relative activities of the signaling proteins C8a, C3a, NF-κB and I-κB, respectively. The functions

$$\text{act}_j(x_i) = \frac{x_i^2}{a_j^2 + x_i^2}, \; j \in \{1, \ldots, 4\}, \tag{3.59}$$

and

$$\text{inh}_j(x_i) = \frac{b_j^2}{b_j^2 + x_i^2}, \; j \in \{1, \ldots, 5\}. \tag{3.60}$$

represent activating and inhibiting interactions. The parameters a_j and b_j are activation and inhibition thresholds, respectively, and take values between 0 and 1. The external TNF stimulus is denoted by u. The initial condition of each single cell is the steady states with C3a = 0 for u = 0. Nominal parameter values are provided in Table 3.1.

It is known from experiments that the cellular response to a TNF stimulus and to other members of the TNF family is highly heterogeneous within a clonal cell population. Some cells die, others survive. This is shown for NCI-H460 cells (non-small lung cancer cells) responding to TNF-related apoptosis inducing ligand (TRAIL) in Figure 3.1 using data from our collaborators, Malgorzata Doszczak and Peter Scheurich (Institute for Cell Biology and Immunology, University of Stuttgart, Germany). The reasons for the heterogeneous behavior are unclear, but of great interest for biological research in TNF signaling, e.g., concerning the use of TNF or related molecules as anti-cancer agents. To understand the biological process

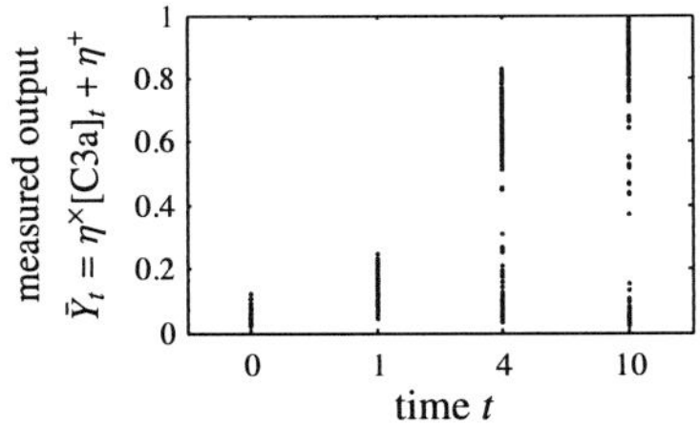

Figure 3.8: Artificial population snapshot data of C3a used to infer the parameter density within the cell population. Each dot (•) represents a single measured cell.

at the physiological and biochemical level it is crucial to consider this cellular heterogeneity, using cell population modeling. While the model used in this study will not provide additional biological insight due it its simplicity, it can provide a proof of concept for the methods.

In the example, we model heterogeneity at the cell level via differences in the parameters b_3 and a_4. The parameter b_3 describes the inhibitory effect of NF-κB via the C3a inhibitor XIAP onto the C3 activity, and the parameter a_4 models the activation of I-κB via NF-κB. Studies showed that these two interactions show large cell-to-cell variability (Albeck *et al.*, 2008a; Paszek *et al.*, 2010; Spencer *et al.*, 2009).

Note that we do not consider stochastic variability, to keep this example short and the discussion simple. A more comprehensive example, also considering stochastic variability, can be found in (Hasenauer *et al.*, 2011c).

Univariate parameter density

For a first evaluation of the proposed method an artificial experimental setup is considered in which the caspase 3 activity is measured at four different time instances during a TNF stimulus,

$$u(t) = \begin{cases} 1 & \text{for } t \in [0, 2] \\ 0 & \text{otherwise.} \end{cases} \tag{3.61}$$

At each time instance the C3a concentration in 150 cells is determined,

$$Y_t = X_{2,t}. \tag{3.62}$$

The concentration measurement is corrupted by additive and multiplicative noise,

$$\bar{Y}_t = \eta^\times Y_t + \eta^+, \tag{3.63}$$

which is realistic in protein quantification (Kreutz *et al.*, 2007). Both components, $\eta^\times$ and η^+, are log-normally distributed (for details see (Hasenauer *et al.*, 2011d)). Median and log-standard deviation are $\mu^\times = 0$ and $\sigma^\times = 0.1$ for $\eta^\times$, and $\mu^+ = \log(0.05)$ and $\sigma^+ = 0.3$ for η^+, respectively. The generated artificial experimental data for a bimodal distribution in b_3, $\theta = b_3$ (Figure 3.10(c)), are depicted in Figure 3.8.

For the parameterization, we employed a generalized gray-box model. The unknown parameter distribution $p_\theta(\theta)$ is parameterized using $d_\varphi = 15$ truncated Gaussians

$$\tilde{\Lambda}_j(b_3) = \begin{cases} \frac{s_j}{\sqrt{2\pi}\sigma} \exp\left(-\frac{(b_3-\mu_j)^2}{2\sigma^2}\right) & b_3 \in [0, 1] \\ 0 & \text{otherwise,} \end{cases} \tag{3.64}$$

where $\sigma = \frac{1}{30}$ and s_j such that $\int_0^1 \tilde{\Lambda}_j(b_3)db_3 = 1$. The center points μ_j are equidistantly distributed on the interval $[0, 1]$. The initial conditions X_0 of the individual cells are the protein levels corresponding to the b_3 dependent steady state with low C3a, namely the surviving state. Thus, we have a functional dependence of X_0 on b_3, which defines the multidimensional ansatz functions $\Lambda(z_0)$.

The prior probability of $p_{\theta,\varphi}(\theta)$ is chosen to be

$$\pi(p_{\theta,\varphi}) = \pi(\varphi) = \begin{cases} \prod_{j=1}^{d_\varphi} p_\beta(\varphi_j|\alpha_j,\beta_j) & \text{for } \mathbf{1}^T\varphi = 1 \\ 0 & \text{otherwise,} \end{cases} \tag{3.65}$$

in which p_β is the probability density of the beta-distribution. The parameters α_j and β_j are selected such that $p_\beta(\varphi_j|\alpha_j,\beta_j)$ has its extremum at $\varphi_{j,\text{ext}} = \frac{1}{s_j}\Big/\left(\sum_{i=1}^{d_\varphi}\frac{1}{s_i}\right)$ and covariance σ^2. Hence, the vector φ which results in the highest prior value results in an almost flat distribution $p_{\theta,\varphi}(\theta)$. The distribution of a sample $\{\varphi^k\}_{k=1}^{d_S}$ drawn from this prior is shown in Figures 3.9(a) and 3.10(a). Note that the prior does not enforce a trend to smaller or larger parameter values of b_3. Furthermore, it does not enforce a trend to unimodal or bimodal distributions $p_{\theta,\varphi}(b_3)$. Such distribution properties shall be inferred from the data.

Given the ansatz functions $\Lambda_j(z_0)$ (3.64) the conditional probabilities densities of observing $\mathcal{D}^i = (\bar{Y}^i_{t^i}, t^i)$, c^i_j, are determined using importance sampling (Algorithm 3.1). This computation takes about three minutes, on a standard personal computer using a single CPU. Thereby, 32% of the computation time are required for the simulation of the single cell model, and 59% for the evaluation of the conditional probability, $p(\bar{Y}^i_{t^i}|Y^i_{t^i})$. The rest is spent on pre- and post-processing. Subsequently, MCMC sampling is performed to obtain a sample $\{\varphi^k\}_{k=1}^{d_S}$ of the prior (with $\sigma^2 = 0.05$), of the conditional, and of the posterior probability distribution. The sample has $d_S = 10^6$ members and the generation takes only four minutes. The computation is very fast, as the proposed approach simplifies the evaluation of the conditional probability to a matrix vector multiplication. Note, that all steps during the computation of the conditional probabilities and the MCMC sampling can be parallelized, yielding a tremendous speed-up for more complex models.

To analyze the information contained in the prior and in the likelihood, and the combined information (posterior), the results of the sampling are illustrated in Figure 3.9 using parallel coordinates (Inselberg & Dimsdale, 1990). From this representation of $\{\varphi^k\}_{k=1}^{d_S}$ it can be seen that after the learning processes most of the density parameters still show large uncertainties. The uncertainty in the posterior distribution is a lot smaller than the uncertainty in the likelihood function, due to the stabilization via the prior, measured in terms of confidence interval width. Note that the visualization also uncovers pronounced correlations between some parameters, e.g., φ_{10} and φ_{11} are negatively correlated for $\varphi^k \sim \mathbb{P}(\mathcal{D}|p_{0,\varphi})$. This indicates that the model of the density of b_3 is over-parameterized with respect to the data. Thus, the number of ansatz functions could be reduced while still achieving a good fit.

To analyze the uncertainty of $p_{\theta,\varphi}(\theta)$ in more detail the sample $\{p_{\theta,\varphi^k}\}_{k=1}^{d_S}$ is employed to determine the 80%, 90%, 95%, and 99% Bayesian confidence intervals. The results are depicted in Figure 3.10. It can be seen that the confidence intervals for some values of b_3 are rather small, indicating that the data contain much information about these regions. Unfortunately, in particular for $b_3 > 0.6$ the confidence intervals are very wide showing that the parameter density in this area cannot be inferred precisely. But, although the amount of data is limited and the uncertainty with single φ_i's may be large, the posterior distribution of $p_{\theta,\varphi}(\theta)$

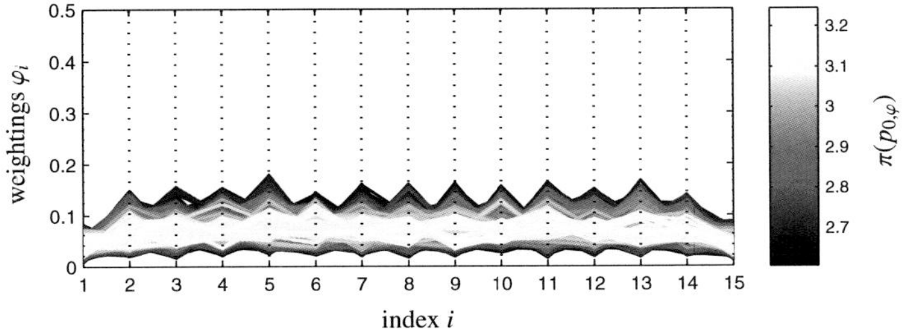

(a) Plot of sample $\{\varphi^k\}_{k=1}^{S_\varphi}$ drawn from prior probability density, $\pi(p_{0,\varphi})$.

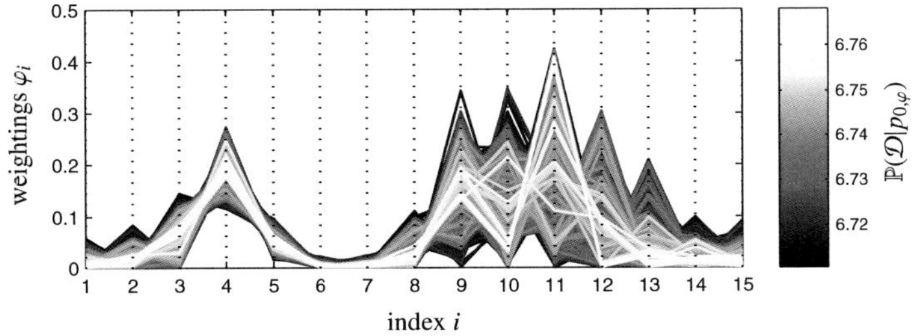

(b) Plot of sample $\{\varphi^k\}_{k=1}^{S_\varphi}$ drawn from conditional probability density, $\mathbb{P}(\mathcal{D}|p_{0,\varphi})$.

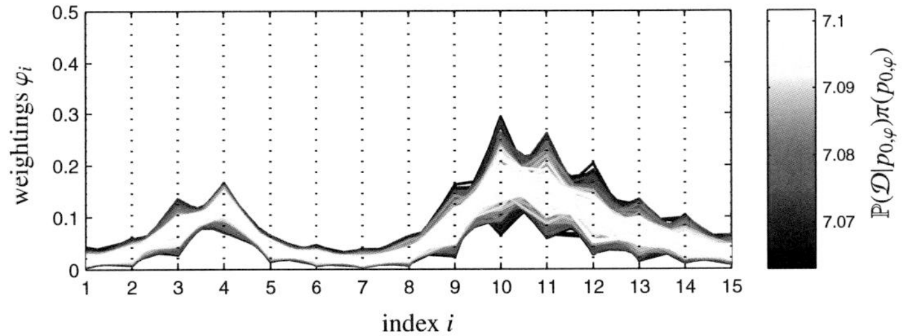

(c) Plot of sample $\{\varphi^k\}_{k=1}^{S_\varphi}$ drawn from posterior probability density, $\pi(p_{0,\varphi}|\mathcal{D})$. As the normalization constant is not computed, the lines are colored with the value of the unnormalized posterior, $\mathbb{P}(\mathcal{D}|p_{0,\varphi})\pi(p_{0,\varphi})$.

Figure 3.9: Visualization of 15-dimensional MCMC sample $\{\varphi^k\}_{k=1}^{d_S}$ from prior, conditional and posterior probability density. Each polyline represents hereby one point φ^k in the 15-dimensional density parameter space. The position of the vertex on the i-th dotted vertical line gives the value of the i-th density parameter. The color of the points indicates the logarithm of the unnormalized probability density of the data. Bright polylines (points φ^k) have a high probability density whereas dark polylines have a low posterior probability.

already shows key properties of the true parameter density, e.g., the bimodal shape, which has not been provided as prior information. This bimodal shape is also seen in the likelihood function, but there the uncertainties are larger than in the posterior probability distribution.

Besides the uncertainty of $p_{\theta,\varphi}(\theta)$ also the predictive power of the model can be evaluated. This is exemplified by studying the confidence interval of the measured C3a and NF-κB concentrations for the previously considered experimental setup. We consider here the noise-corrupted outputs, as this allows for the direct comparison of the prediction and artificial measurement data used for parameter estimation. The predictions are shown in Figure 3.11.

It is obvious that, although the parameter distributions show large uncertainties, the predictions are rather certain. This is indicated by tight confidence intervals. Furthermore, the mean predictions $\mathbb{E}\left[p(\bar{y}_{i,t}|t, p_{0,\varphi})\right]$ and the predictions with highest posterior probability $p(\bar{y}_{i,t}|t, p_{0,\varphi^*})$ agree well with the true output distribution $\mathbb{E}\left[p(\bar{y}_{i,t}|t, p_0^{\text{true}})\right]$, for measured output C3a and predicted output NF-κB. The small prediction uncertainties can be explained to be sloppiness (Apgar *et al.*, 2010; Gutenkunst *et al.*, 2007) of the density parameters φ_i parametrizing the distribution of b_3. A more detailed analysis indicates that the number of ansatz function can be decreased, still ensuring a good approximation of the distribution of b_3.

Multivariate parameter density

In many biological systems several cellular parameters are heterogeneous and different cellular concentrations can be measured (Albeck *et al.*, 2008a). Therefore, we show in this section that the proposed method can also be employed to estimate multivariate parameter densities from multi-dimensional outputs. Furthermore, the influence of the choice of the prior on the estimation result is analyzed.

To perform this study we considered the same experimental setup as above. The only differences are that two concentrations are measured, C3a and NF-κB, $Y_t = [X_{2,t}, X_{3,t}]^{\text{T}}$, and that the joint distribution of $\theta = [a_4, b_3]^{\text{T}}$ is estimated. The considered artificial experimental data of 10^4 cells are depicted in Figure 3.12. The ansatz function for $p_\theta(\theta)$ are $n_\varphi = 100$ truncated multivariate Gaussians equivalently to (3.64). The covariance matrix is $0.06^2 \cdot \text{I}_2$ and the extrema are equidistantly distributed on a regular 2-dimensional grid which is aligned with the axes.

Given this setup, the convergence rate is studied in terms of the integrated mean square error,

$$\text{IMSE} = \int_{[0,1]^2} \left(p_\theta^{\text{true}}([a_4, b_3]^{\text{T}}) - p_{\theta,\varphi^*}([a_4, b_3]^{\text{T}})\right)^2 da_4 db_3, \tag{3.66}$$

of the true distribution, $p_\theta^{\text{true}}([a_4, b_3]^{\text{T}})$, and the distribution with the highest posterior probability, $p_{\theta,\varphi^*}([a_4, b_3]^{\text{T}})$. The IMSE is computed for different amounts of measured cells per time instance and different priors. The priors are thereby again beta-distributions (3.65). The extrema φ_{ext} are chosen as in the last section such that the prior is flat. The standard deviation on the other hand is reduced step-wise from $\sigma = 0.285$ (completely uninformative as almost uniform on the feasible interval $\varphi \in [0, 1]^{d_\varphi}$) to $\sigma = 0.001$ (very informative). Given these requirements, the values α_i and β_i of the prior (3.65) are determined. The result for different numbers of measured cells sampled from the available data set is shown in Figure 3.13. Note that the IMSE is a stochastic quantity as the selection of measured cells is a stochastic processes and hence also the estimated density $p_{\theta,\varphi^*}([a_4, b_3]^{\text{T}})$ is stochastic. To account for this stochasticity, several realizations are performed and the mean is computed.

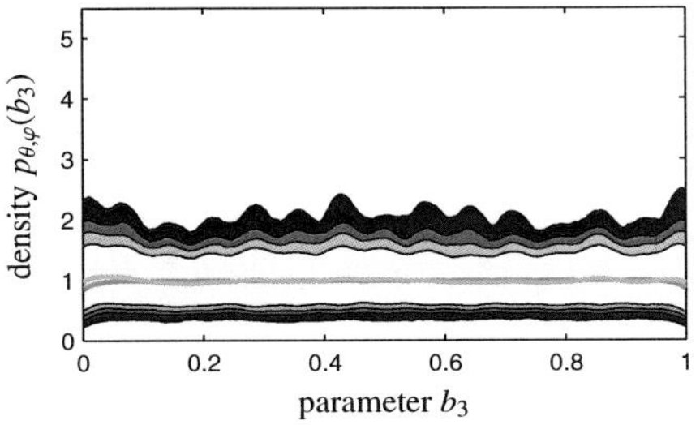

(a) Prior probability density, $\pi(p_{\theta,\varphi})$.

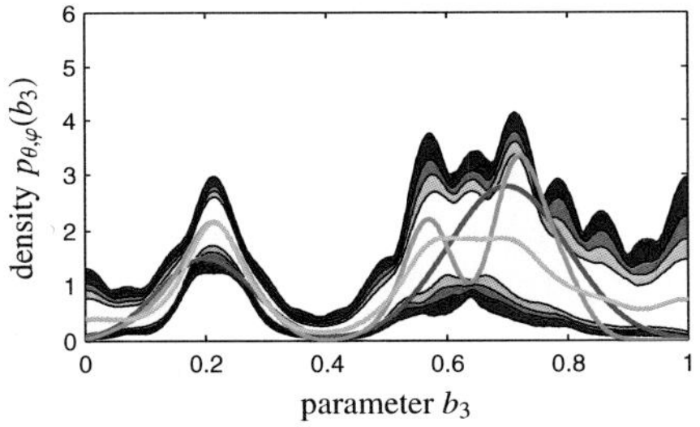

(b) Conditional probability density, $\mathbb{P}(\mathcal{D}|p_{\theta,\varphi})$.

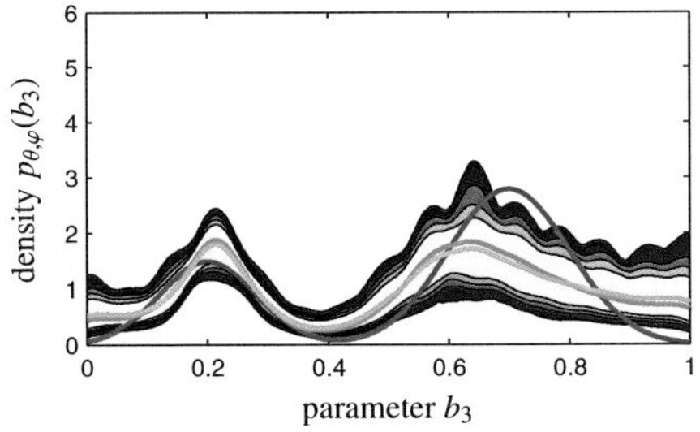

(c) Posterior probability density, $\pi(p_{\theta,\varphi}|\mathcal{D})$.

Figure 3.10: Prior, conditional and posterior probability density of $p_{\theta,\varphi}(b_3)$ in b_3 - $p_{\theta,\varphi}(b_3)$ - plane. The colored lines indicate the distribution with the highest posterior probability $p_{\theta,\varphi^*}(b_3)$ (—), and the mean distribution $\mathrm{E}[p_{\theta,\varphi}](b_3)$ (—), for the individual probability densities, as well as the true parameter distribution $p_\theta^{\mathrm{true}}(b_3)$ (—). The colored regions indicate the 80% ( ), 90% (■), 95% (■), and 99% (■) Bayesian confidence intervals of the parameter distribution $p_{\theta,\varphi}$.

Legend:

— maximum a posterior density, $p(\bar{y}|t, p_{0,\varphi^*})$

— mean density, $\mathbb{E}\left[p(\bar{y}|t, p_{0,\varphi})\right]$

— density used for data generation, $p(\bar{y}|t, p_0^{\text{true}})$

• measurement data, $\bar{Y}_t$

■ 80% confidence intervals of $p(\bar{y}|t, p_{0,\varphi})$

■ 90% confidence intervals of $p(\bar{y}|t, p_{0,\varphi})$

■ 95% confidence intervals of $p(\bar{y}|t, p_{0,\varphi})$

■ 99% confidence intervals of $p(\bar{y}|t, p_{0,\varphi})$

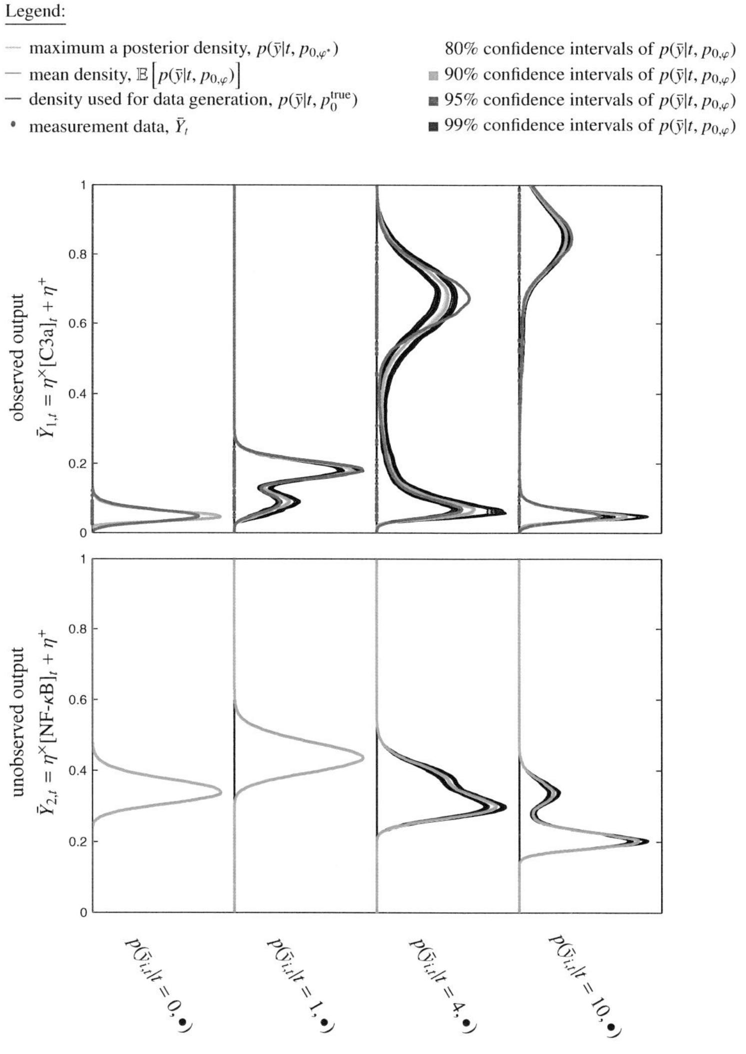

Figure 3.11: Predicted density $p(\bar{y}_{i,t}|t, p_{0,\varphi})$ of the measured [C3a] and [NF-κB] concentrations. The colored lines indicate the distribution with the highest posterior probability $p(\bar{y}_{i,t}|t, p_{0,\varphi^*})$ (—), and the mean distribution $\mathbb{E}\left[p(\bar{y}_{i,t}|t, p_{0,\varphi})\right]$ (—), as well as the true measured output distribution $p(\bar{y}_{i,t}|t, p_0^{\text{true}})$ (—). The colored regions indicate the 80% ( ), 90% (■), 95% (■), and 99% (■) Bayesian confidence intervals of the predicted distribution $p(\bar{y}_{i,t}|t, p_{0,\varphi})$.

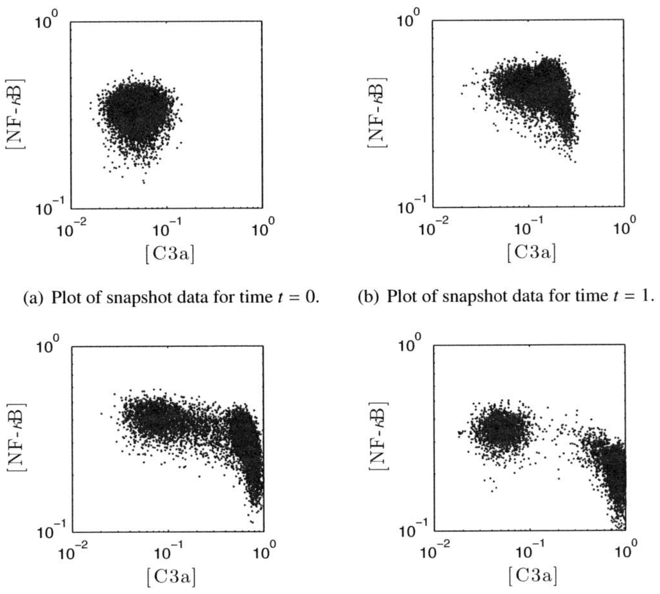

(a) Plot of snapshot data for time $t = 0$.

(b) Plot of snapshot data for time $t = 1$.

(c) Plot of snapshot data for time $t = 4$.

(d) Plot of snapshot data for time $t = 10$.

Figure 3.12: Artificial population snapshot data of C3a and NF-κB used to infer the parameter density within the cell population. Each dot (•) represents a single measured cell.

From Figure 3.13 it becomes clear that the IMSE strongly depends on both, amount of data and informativeness of the prior. For uninformative priors, the outcome for small amounts of data is highly uncertain and the IMSE is large and shows large variations. However, if the prior is very informative but wrong, the number of measurement data required to obtain a good estimate is tremendous. For the right choice of σ, one observes a fast convergence of the $p_{\theta,\varphi^*}([a_4, b_3]^T)$ to $p_\theta^{\text{true}}([a_4, b_3]^T)$, as shown in Figure 3.14, and only little variation for small amounts of data. Hence, the usage of prior knowledge, even if it is only partially correct, yields more stable estimates and faster convergence. Furthermore, this study suggests that the typical number of cells measured by flow cytometry (10^4) is sufficient to infer key features of population heterogeneity.

3.5 Summary and discussion

In this chapter, we studied the problems of modeling (Problem 3.1) and parameter estimation (Problem 3.2) for signal transduction in cell populations exhibiting intrinsic and extrinsic noise. Therefore, we introduced in Section 3.2 a model for the time evolution of the state and output density for heterogeneous cell populations. To obtain the population model, the

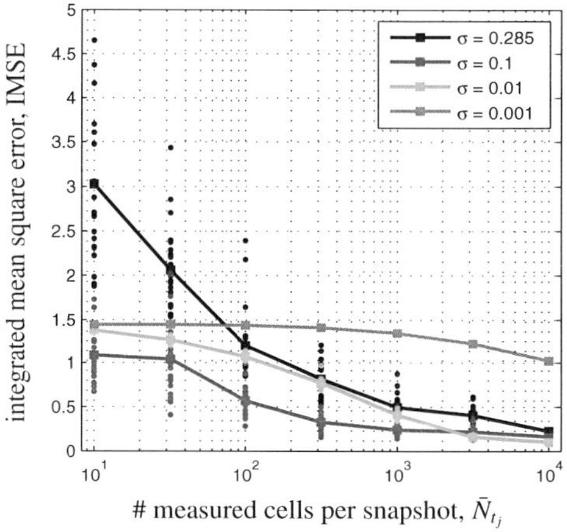

Figure 3.13: Integrated mean square error as function of the amount of available data and the informativeness of the prior. The plot shows the integrated mean square error for different numbers of measured cells per time instance and different standard deviation, σ, of the prior. Individual realization (•) and resulting mean (–▪–).

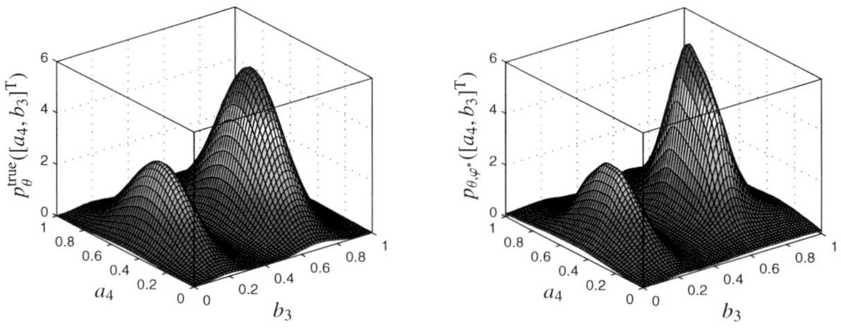

(a) True parameter density, $p_\theta^{\mathrm{true}}(\theta)$.

(b) Estimated parameter density, $p_{\theta,\varphi^*}(\theta)$.

Figure 3.14: Estimation result for 2-dimensional parameter density. The estimated parameter density is obtained using 10^4 measured cells at each time instance and a prior with $\sigma = 0.01$.

stochastic and parameter dependent single cell model is transformed using an augmented state vector which contains the actual state of the cell and its parameters, $Z_t = [X_t^T, \theta^T]^T$. The probability distribution of the state Z_t of this augmented single cell model, $p(z|t, p_0)$, is described by an augmented Fokker-Planck equation, while the output density, $p(y|t, p_0)$, can be computed via marginalization. To simulate this population model, a particle-based approach is combined with kernel density estimation, allowing for the efficient assessment and prediction of population heterogeneity. The augmented Fokker-Planck equation and the introduced simulation algorithm provides a solution to Problem 3.1.

Based upon the population model, we posed in Section 3.3 the problem of inferring the population heterogeneity from single cell snapshot data. To infer the population-wide distribution of parameters and initial states, $p_0([x_0^T, \theta^T]^T)$, a Bayesian framework is used, which relies on an affine parametrization of $p_0([x_0^T, \theta^T]^T)$. The affine parameterization enables the reformulation of the likelihood, which gives rise to a two-step procedure. In the first step, a parametric form of the likelihood is computed which can be evaluated efficiently, without any need for simulation. In the second step, this parametric form of the likelihood is used to compute the maximum a posteriori estimate, and to generate a representative sample from the posterior distribution.

The decomposition of the parameter estimation in (1) the precomputation of the parametric form of the likelihood and (2) the optimization and uncertainty analysis allows for an efficient overall procedure. The maximum a posteriori estimate can in most cases even be computed as solution of a quasiconvex optimization problem. To determine parameter and prediction uncertainties, we generated a sample from the posterior probability using adaptive Metropolis-Hastings sampling with delayed rejection (Haario *et al.*, 2006). This sampling approach is adapted to be applicable to the considered constraint problem.

The properties of the proposed scheme are evaluated using artificial data of a TNF signal transduction model. It could be shown that the proposed method yields good estimation results for a realistic experimental setup. Furthermore, although the remaining uncertainties are large, the predictions show only small uncertainty indicating sloppiness of parameters. Hence, over-parameterization of the distribution of parameters and initial states might be an issue, but information on the output levels are correctly reproduced. Concerning the choice of the prior distribution it could be shown that the regularizing effect is beneficial if only little data is available. On the other hand, if the amount of available data increases, informative but not carefully chosen priors slow down the convergence. While in this example we considered deterministic single cell dynamics for the benefit of clarity, a study with stochastic single-cell dynamics can be found in (Hasenauer *et al.*, 2010a).

The Bayesian approach presented in this chapter is the result of an extensive development process, during which several other methods have been studied by the authors. Alternatives to the Bayesian approach discussed in this thesis are a heuristic sampling-based approach (Waldherr *et al.*, 2009), a density-based approach (Hasenauer *et al.*, 2011c, 2010a), and a frequentist approach (Hasenauer *et al.*, 2010b). The earliest method, the sampling-based approach (Waldherr *et al.*, 2009), relies on an entropy consideration and performs well for noise-free data. To account for measurement noise, density-based approaches have been introduced (Hasenauer *et al.*, 2011c, 2010a), which employ a density model of the measured output distribution derived from extensive single cell measurements. Given this density model of the data and the population model, a l_2-norm minimization can be performed. To reduce the required amount of single cell measurements, a likelihood function has been derived for population snapshot data, which allows for a frequentist approach (Hasenauer *et al.*, 2010b). Still,

Table 3.2: Properties of the published estimation methods, allowing for the inference of parameter distributions in heterogeneous cell populations from single cell snapshot data.

method	required amount of data	measurement noise	consideration of prior information	parameter uncertainties	prediction uncertainties	stochastic single cell models	statistical interpretation
heuristic sampling-based approach (Waldherr *et al.*, 2009)	large	–	–	–	–	–	–
density-based approach (Hasenauer *et al.*, 2011c, 2010a)	large	✓	–	✓	–	✓	–
frequentist approach (Hasenauer *et al.*, 2010b)	moderate	✓	–	–	–	–	✓
Bayesian estimation approach (Hasenauer *et al.* 2011d, this thesis)	small	✓	✓	✓	✓	✓	✓

neither prior information could be incorporated nor global uncertainty bounds were available. This motivated the development of the Bayesian methods which outperformed all previously available approaches (for details see Table 3.2).

Note that the methods developed in this thesis have to be considered as the first modeling and parameter estimation methods for heterogeneous cell populations which can be applied to incorporate single cell snapshot data. Thus, our new method complements tools which employ single cell time-lapse data (Kalita *et al.*, 2011; Koeppl *et al.*, 2012). In addition, the method extends results obtained for cell populations exhibiting purely stochastic cell-to-cell variability, as presented in (Munsky & Khammash, 2010; Munsky *et al.*, 2009; Nüesch, 2010). Also note that the presented methods can be extended to single cell dynamics governed by Markov jump processes.

To sum up, the extension of available methods to infer the deterministic/regulated population heterogeneity from vast amounts of single cell snapshot data is nontrivial. The main challenge is the computational complexity of simulating the model and evaluating the likelihood function. In this chapter, we presented the first method to overcome this problem, employing an efficient formulation of the problem, which allows the decomposition of the estimation process. This enables an in-depth analysis of population dynamics of heterogeneous cell systems, as well as the source of heterogeneity.

4 Proliferation of heterogeneous cell populations

In the previous chapter, we analyzed the signal transduction in heterogeneous cell populations. To complement this work, we study the genesis of populations and provide methods to assess proliferation properties in this chapter. In Section 4.1, we provide a formal introduction to the considered process and measurement data, and state the problems. Based upon this, we derive in Section 4.2 a division- and label-structured population (DLSP) model, which accounts for heterogeneity in different cellular properties. In Section 4.3, we introduced an approach to estimate the parameters of the DLSP model from proliferation assay data. For this, we provide a likelihood function as well as an approximation of this likelihood function. Subsequently, we show how simulation and likelihood function evaluation can be intertwined, yielding an efficient parameter estimation method. The proposed approach is applied to data from T lymphocytes in Section 4.4. The results are summarized and discussed in Section 4.5.

The section on modeling of cell proliferating in based upon (Hasenauer *et al.*, 2012c), while a part of the section on parameter estimation is based upon (Hasenauer *et al.*, 2011b).

4.1 Introduction and problem statement

4.1.1 Biological background

Cell proliferation is a central aspect of most biological processes, among others bacterial growth (Gyllenberg, 1986; Zwietering *et al.*, 1990), immune response (De Boer *et al.*, 2006; Hodgkin *et al.*, 1996; Luzyanina *et al.*, 2007a), stem cell induced tissue remodeling (Buske *et al.*, 2011; Glauche *et al.*, 2009), and cancer progression (Anderson & Quaranta, 2008; Eissing *et al.*, 2011). Depending on the biological process, cellular proliferation has different characteristics. Cell division can be symmetric or asymmetric, and the daughter cells may or may not inherit the age of the mother cells (see Figure 4.1). While many micro-organisms, such as the budding yeast, grow a daughter cell (Shcheprova *et al.*, 2008) which does not inherit the age of the mother cell and is born young, in most multicellular organisms the mother cell divides symmetrically into two daughter cells which inherit the age of the mother cell (Hayflick, 1965). The latter proliferation type results in an accumulation of DNA damage and telomere shortening, which may be interpreted as aging of the individual cell. This manifests in a reduced proliferation potential, a reduced proliferation speed and finally in cell cycle arrest (Glauche *et al.*, 2011; Hayflick, 1965, 1979), known as senescence (Gewirtz *et al.*, 2007). This has been discovered by Hayflick (1965) in the 1960s and the upper limit for the number of cell divisions a normal cell can undergo has been termed Hayflick limit.

While aging of individual cells results in aging of the cell population, cell populations do not age homogeneously. It is well-known that cell cycle progression depends on many factors, e.g., the abundance of transcription factors and the partitioning at cell division (Avery,

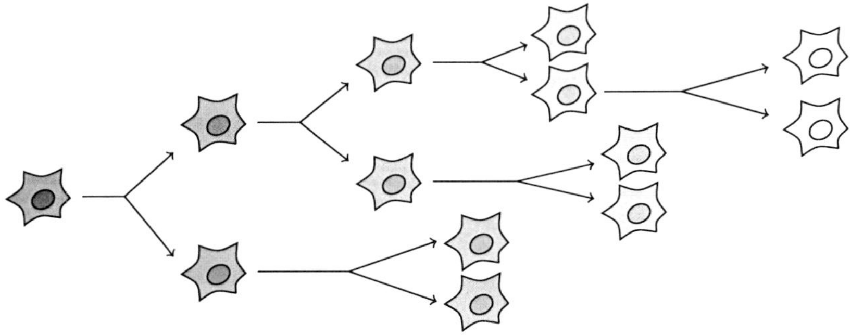

(a) Mother cells undergoing symmetric cell division cells split up into almost identical daughter cells. Differences between daughter cells are merely caused by the stochasticity of DNA replication, resulting in genetic and epigenetic differences, and the stochasticity of cell partitioning, yielding different protein abundances.

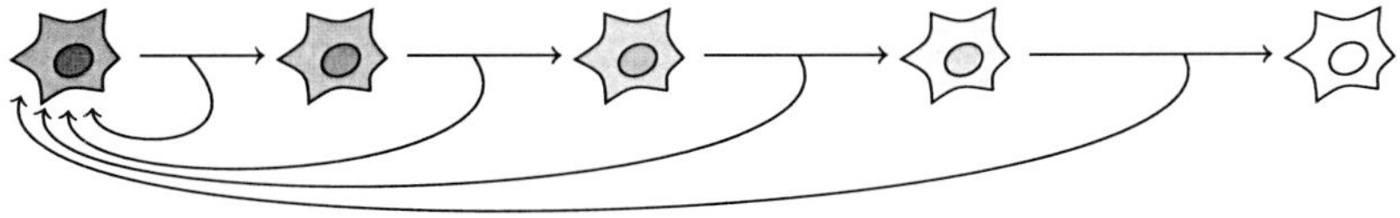

(b) Mother cells undergoing asymmetric cell division yield daughter cells with different cell fates. In the most extreme case, the mother cells grows a daughter cell – this is called budding. The daughter cell is genetically identical to the mother cell but is born young, meaning that it does not inherit the age of the mother cell.

Figure 4.1: Illustration of symmetric and asymmetric cell division. The color of the cells indicate the cell's age. In case of symmetric cell division the age strongly correlates with the number of divisions a cell has undergone.

2006; Hilfinger & Paulsson, 2011), and is a stochastic process. Different cells do not divide after exactly the same period of time, some divide early while others divide late. Therefore, cells within a cell population are of different age, measured in terms of cell divisions, due to differences in their history. These differences result in cell-to-cell variability (Bird *et al.*, 1998; Hodgkin *et al.*, 1996) which is known to increase the fitness of the population (Avery, 2006).

In this thesis, we focus on symmetric division processes and heterogeneity arising due to differences in the number of cell divisions. This is a topic of active research in immunology, as well as stem cell biology.

4.1.2 Proliferation assays

Experimentally, proliferation dynamics of cell populations are mainly analyzed using proliferation assays. These proliferation assays employ molecules which are normally not found in cells, such as Bromodeoxyuridine (BrdU) (Gratzner, 1982) and Carboxyfluorescein succinimidyl ester (CFSE) (Lyons & Parish, 1994). As mainly CFSE is used in recent studies, we will focus on CFSE but the methods directly extend to BrdU labeling.

Labeling with CFSE

To study the proliferation dynamics of cells using CFSE, the cells are incubated with carboxyfluorescein diacetate succinimidyl ester (CFDA-SE). CFDA-SE is highly similar to CFSE, it merely possesses additional acetate groups. Due to these acetate groups, CFDA-SE is highly cell permeable and can cross the cell membrane. In the cytoplasm of the cells, the acetate groups are removed by intracellular esterases, thereby converting the non-fluorescent CFDA-SE into the fluorescent CFSE. Finally, CFSE couples covalently to intracellular proteins via its succinimidyl group. The labeling process is illustrated in Figure 4.2.

CFSE is a fluorescent cell staining dye which stays in cells for a long time. Its concentration is in the following denoted by $x \in \mathbb{R}^+$. This concentration reduces only due to protein degradation and cell division (see in-depth discussion in Section 4.2.1). At cell division, CFSE is distributed approximately equally among daughter cells (Lyons & Parish, 1994). Thus, the proliferation of labeled cells results in a progressive dilution of the dye (Luzyanina *et al.*, 2009), as depicted in Figure 4.3.

Although most key properties of CFSE labeling and proliferation assays in general are understood, some questions remain open. In particular, there are several competing models for the CFSE degradation. While some studies assume a time invariant degradation process (Banks *et al.*, 2010; Luzyanina *et al.*, 2009, 2007b), others employ Gompertz decay processes (Banks *et al.*, 2011), where the degradation rate decreases with time. Experimentally, the latter seems to describe the label degradation better, but to our best knowledge a valid mathematical argument is still missing.

Single cell measurement of CFSE intensity

To obtain quantitative information about the proliferation dynamics, the fluorescent levels of individual cells are assessed using flow cytometry (Hawkins *et al.*, 2007). These fluorescence level of an individual cell, $\bar{Y} \in \mathbb{R}_+$, summarizes the CFSE induced fluorescence, $Y = cX \in \mathbb{R}_+$, and the autofluorescence, $Y_a \in \mathbb{R}_+$,

$$\bar{Y} = Y + Y_a = cX + Y_a. \tag{4.1}$$

The CFSE induced fluorescence is proportional to the amount of CFSE, X, where $c \in \mathbb{R}_+$ is the fluorescence intensity per CFSE molecule. The autofluorescence – also called background fluorescence – is a stochastic variable, which is independent of the level of CFSE concentration. While the CFSE concentration, X, and therefore the CFSE induced fluorescence, Y, halves at cell division, this is not true for the autofluorescence.

Experiments showed that the autofluorescence varies among cells (Hawkins *et al.*, 2007), $Y_a \sim p_a(y_a)$. In many experiments it is approximately log-normally distributed (Banks *et al.*, 2012; Hawkins *et al.*, 2007; Thompson, 2012),

$$p_a(y_a) = \log \mathcal{N}(y_a|\mu_a, \sigma_a^2), \tag{4.2}$$

with parameters $\mu_a \in \mathbb{R}$ and $\sigma_a^2 \in \mathbb{R}_+$. This autofluorescence, which might be interpreted as measurement noise, avoids a precise reconstruction of the CFSE concentration. Furthermore, it limits the number of cell divisions which can be observed. As the initial CFSE concentration cannot be arbitrarily high to avoid interference with the cell's functionality

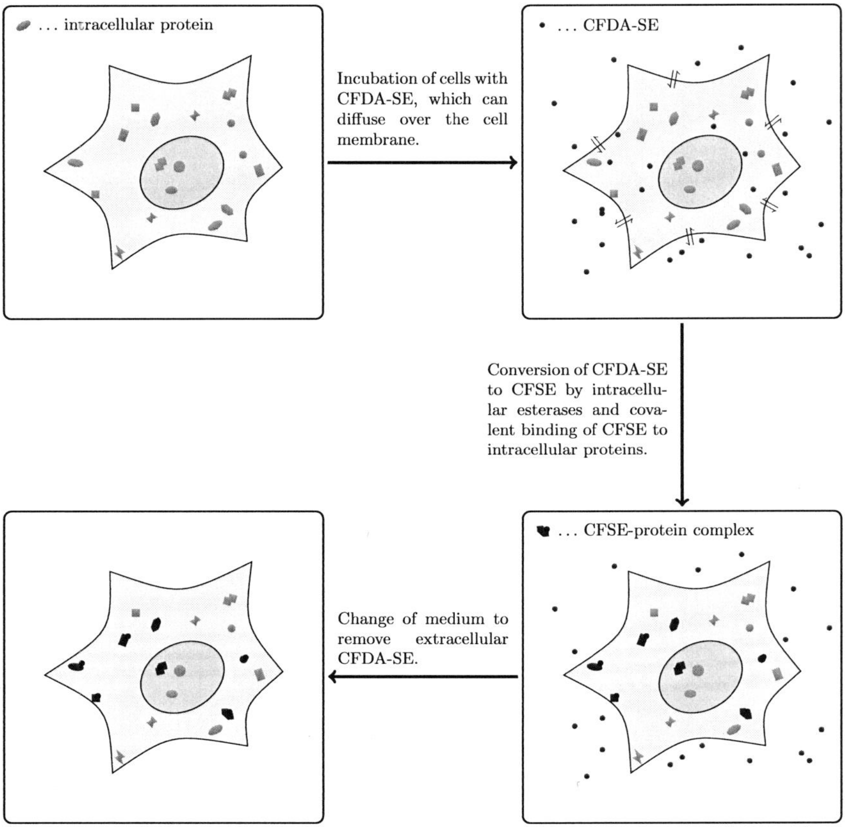

Figure 4.2: Schematic of the CFSE labeling process, including the incubation of cells with CFDA-SE, the diffusion of CFDA-SE through the cell membrane, the conversion of CFDA-SE into CFSE and the covalent binding of CFSE to intracellular proteins. As CFSE is bound covalently to proteins, its concentration only decreases due to protein degradation and cell division.

and toxicity, even in case of highly optimized labeling strategies only six to eight division can be observed before the observed fluorescence is indistinguishable from the background fluorescence (Hawkins *et al.*, 2007).

Binned snapshot data

Flow cytometry allows for the automated assessment of large numbers of cells, in general 10^3 – 10^6 individual cells per snapshot (Hawkins *et al.*, 2007; Hodgkin *et al.*, 1996; Lyons & Parish, 1994). Due to the often large number of measured cells, most flow cytometers directly bin the data (Lampariello & Aiello, 1998; Overton, 1988). This binning, also called

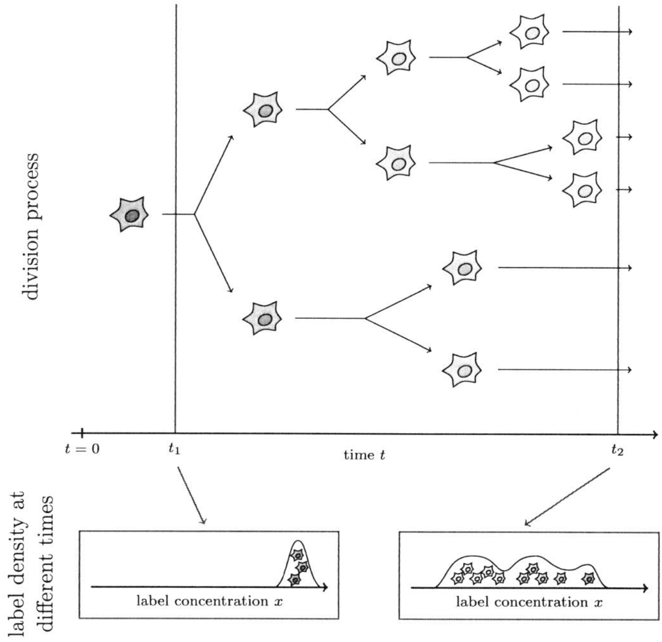

Figure 4.3: Illustration of CFSE dilution due to cell division. The division process results in halving of the concentration at each division (top), yielding a multimodal label intensity distribution within the cell populations (bottom). As cells do neither divide at the same time nor proliferate at the same speed, several modes might be observed. These modes belong to different division numbers. Depending on the labeling, the modes can show large overlaps (see discussion in Hawkins *et al.* (2007)).

channeling, is a lossy data compression. The obtained data are in the following denoted as binned snapshot data, $\mathcal{SD}^{\mathcal{B}}$.

We emphasize that not all information about the individual cells can be reconstructed from the binned snapshot data. Therefore, single cell snapshot data are more informative. Still, if the bin width is small, large portions of the single cell information is available, as an interval bound for the output of each measured cell is known.

Experimentally, binned snapshot data are obtained by drawing $\bar{N}_{t_j}$ cells from the cell population and measuring one (or more properties) of these cells (here the fluorescence intensity y). This yields the single cell snapshots $\mathcal{SD}_j = \left\{\left(\bar{Y}^i_{t_j}, t_j\right)\right\}_{i=1}^{\bar{N}_{t_j}}$, with noise-corrupted output $\bar{Y}^i_{t_j}$, as introduced earlier. To obtain binned snapshot data, additionally a set of intervals/bins $I_l = [\bar{y}_{l,\mathrm{lb}}, \bar{y}_{l,\mathrm{ub}})$, $l = 1, \ldots, d_I$, is defined or specified by the measurement device, respectively. Lower and upper bounds of the interval are denoted by $\bar{y}_{l,\mathrm{lb}}$ and $\bar{y}_{l,\mathrm{ub}}$, respectively. Lower and upper bounds are non-decreasing in l, with $\bar{y}_{l+1,\mathrm{lb}} = \bar{y}_{l,\mathrm{ub}}$, $\bar{y}_{1,\mathrm{lb}} = 0$, and $\bar{y}_{d_I,\mathrm{ub}} = \infty$. Given the

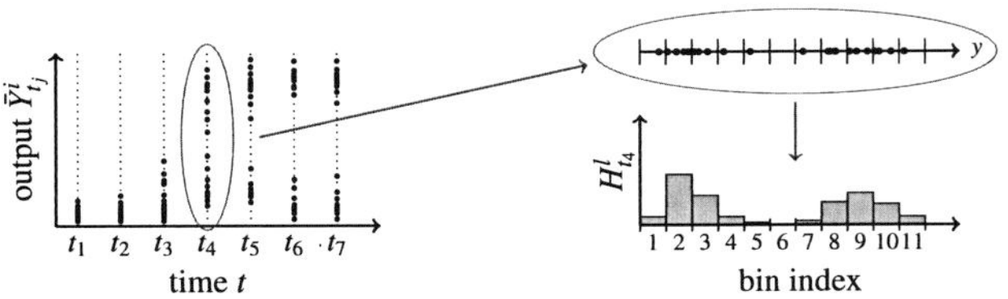

Figure 4.4: Illustration of the collection process of binned snapshot data. The individual steps are: (1) gathering of snapshot data; (2) assignment of the single cell data to bins; and (3) counting of the number of cells per bin.

intervals I_l, $i = 1, \ldots, d_l$, the number of cells with measured output $\bar{Y}^i_{t_j} \in I_l$ are determined,

$$\bar{H}^l_{t_j} = \text{card}\left(\{\bar{Y}^i_{t_j}, i \in \{1, \ldots, \bar{N}_{t_j}\} | \bar{Y}^i_{t_j} \in I_l\}\right). \tag{4.3}$$

Employing $\bar{H}^l_{t_j}$, the binned snapshot data are

$$\mathcal{D} = \bigcup_{j=1}^{d_{\mathcal{D}}} \mathcal{SD}^{\mathcal{B}}_j, \quad \text{with} \quad \mathcal{SD}^{\mathcal{B}}_j = \left\{\left(\bar{H}^l_{t_j}, t_j\right)\right\}_{l=1}^{d_l}. \tag{4.4}$$

The collection process of binned snapshot data, which are simply unnormalized histograms (Silverman, 1986), is sketched in Figure 4.4.

4.1.3 Available methods to analyze CFSE data

Proliferation assays are mainly used to analyze the rates of cell division and of cell death, and their dependence on time, division number, and external stimuli (see, e.g., (Banks *et al.*, 2011; Luzyanina *et al.*, 2007b)). To determine these proliferation properties of cells from CFSE data, several analysis tools have been introduced.

The methods proposed first are pure data analysis tools. They employ peak detection and deconvolution (Hawkins *et al.*, 2007; Luzyanina *et al.*, 2007a; Nordon *et al.*, 1999) to determine the number of cells with a certain division number. Unfortunately, these data analysis methods are only applicable if the modes, corresponding to cells with a common division number, are well separated and if the data are not strongly noise corrupted. As discussed above, if several divisions are observed, these requirements are in general not met due to cellular autofluorescence. To overcome these limitations, different model-based approaches have been developed.

In the literature, mainly three different classes of population models are described: exponential growth models, division-structured population models and label-structured population models. The exponential growth (EG) models are the simplest ones, and merely describe the number of individuals in a cell population. For this task a one-dimensional ODE, like the Gompertz equation (Zwietering *et al.*, 1990), is sufficient. While exponential growth models allow for the description of the proliferation of many bacterial populations, they are in general not capable of describing the dynamics of human tissue cells. One reason for this is that for many cell systems the cell division and cell death rates are known to depend on the division number (Hayflick, 1965), e.g., B cells (De Boer *et al.*, 2006), T cells (De Boer *et al.*, 2006), and osteoblasts (Kassem *et al.*, 1997).

To capture division number-dependent effects, a multitude of division-structured population (DSP) models has been introduced (De Boer *et al.*, 2006; Deenick *et al.*, 2003; Lee & Perelson, 2008; León *et al.*, 2004; Nordon *et al.*, 1999; Revy *et al.*, 2001; Yates *et al.*, 2007). The state variables of these models describe the sizes of the subpopulations, which are defined by a common division number. Hence, these models allow for the consideration of division number-dependent properties. Still, these models do not provide information about the label concentration and thus cannot be compared to data directly but require complicated and error-prone data processing.

To avoid this, label-structured population models (LSP) are employed (Luzyanina *et al.*, 2007b). These models describe the evolution of the population density on the basis of an one-dimensional hyperbolic PDE. Hence, they provide predictions for the label distributions at the individual time points and can even take autofluorescence into account (Banks *et al.*, 2011). This allows for the fitting of label-structured population to CFSE data (Banks *et al.*, 2011, 2010; Luzyanina *et al.*, 2009, 2007b), and thereby for the assessment of proliferation rates. This renders complex data processing redundant and simplifies the model-data comparison. Still, these models do not allow for a direct consideration of division number-dependent parameters. Complex dependencies of the cell division and cell death rates on time and label concentration have been introduced (Banks *et al.*, 2011, 2010). However, these are neither intuitive nor easy to interpret. Furthermore, the simulation of label-structured population models is computationally demanding and requires discretization. Finally, parameter estimation for label-structured population models mainly employs norm-distances between model prediction and measured data. The resulting estimates therefore lack statistical interpretation and their reliability is not clear.

4.1.4 Problem formulation

As outlined above, there are several challenges associated to the analysis of proliferating cell populations, employing CFSE based proliferation assays. These challenges concern the modeling of individual cells, as well as modeling of cell populations.

Problem 4.1. (Modeling of CFSE dynamics in labeled individual cells) *Given the binding properties of CFSE to intracellular proteins, describe the time evolution of the label concentration in individual cells.*

Problem 4.2. (Modeling of CFSE labeled proliferating cell populations) *Given the CFSE dynamics in individual cells, describe the time evolution of proliferating cell populations, accounting for the label dynamics and the heterogeneity induced by varying division numbers.*

To approach Problem 4.1, the degradation process will be studied carefully, accounting for varying half-life times of CFSE bound molecules. Problem 4.2 will be approached by combining division-structured population models and the label-structured population models and thereby overcomes their individual shortcomings. This population model, which we termed division- and label-structured population (DLSP) model, is based on our own work (Hasenauer *et al.*, 2012c; Schittler *et al.*, 2011).

In addition, the developed models have to be linked to available data to determine the proliferation properties of the cell population.

Problem 4.3. (Estimation and uncertainty analysis of proliferation parameter) *Given the binned snapshot data $\mathcal{SD}^{\mathcal{B}}$, the model for autofluorescence* (4.2)*, and a model for the dynamics of CFSE labeled cell population, determine the unknown division and death rates, and their uncertainty.*

In this work, we will employ Bayesian methods to infer the parameters and to evaluate their parameter uncertainties. As Bayesian methods perform a global exploration of the parameter space, an efficient evaluation procedure for the likelihood is necessary. Such a procedure is proposed, based on the linkage of simulation and likelihood function evaluation.

4.2 Modeling the proliferation of heterogeneous cell populations

In this section, we approach Problem 4.1 and 4.2. Therefore, we analyze in Section 4.2.1 the overall degradation rate in systems composed of a variety of species. This allows for further insight into the process of CFSE degradation. In addition, a division- and label-structured population model is introduced in Section 4.2.2 and incorporates both aspects: Discrete changes of the cell division number due to cell divisions, and continuous dynamics of the label distribution. The overall model is a system of coupled partial differential equations. We discuss how this system of PDEs can be split up into two decoupled parts in Section 4.2.3, namely a single PDE and a set of ODEs, which significantly simplifies the solution. The obtained model is reduced further by truncation of the state space. This truncation and the resulting truncation error can be controlled using the a priori error bound which we derive. As the proposed model unifies the existing models, we outline the relationship of the models in Section 4.2.4.

4.2.1 Modeling the CFSE decay process

As discussed earlier, the CFSE concentration is reduced by two mechanisms: cell division and label degradation. While the former is well understood – CFSE is divided approximately equally among the daughter cells (see, e.g., discussion in (Hawkins *et al.*, 2007; Lyons & Parish, 1994)) –, only few studies even discuss the degradation dynamics. The reason might be that in most biological studies, the degradation is not of interest, as merely peak detection is performed to analyze the data (see, e.g., (Hawkins *et al.*, 2007; Hodgkin *et al.*, 1996; Luzyanina *et al.*, 2007a; Lyons & Parish, 1994; Lyons, 2000; Nordon *et al.*, 1999)). However, sophisticated model-based methods rely on a description of the degradation process. Such description can be derived from experimental data or from the mechanisms of the underlying biochemical processes. Using the former approach, it was shown by Banks *et al.* (2011) that the Gompertz decay process provides a better model of the label dynamics than common time independent decay processes. In the following, a model-based approach will be used to support this finding and to provide a mechanistic understanding.

To unravel the source of the time dependence and to propose better models, we study the degradation process in more detail. Therefore, we assume that (1) CFSE is degraded if the protein to which it is bound is degraded, and (2) the degradation of individual proteins is a linear and time independent process. While the first assumption is consistent with current biological understanding as CFSE is covalently coupled to the protein via its succinimidyl group, a variety of degradation mechanisms exists. Therefore, the second assumption might not hold for all proteins, but the assumption is reasonable to gain first insights into the process.

Given the previous assumption, let $P(S_i)$ be the probability that CFSE binds to any protein of species S_i, $i = 1, \dots, d_S$. The degradation rate of the complex of S_i and CFSE is $k_i \in \mathbb{R}_+$, yielding

$$\forall i : S_i\sim\text{CFSE} \xrightarrow{k_i} \emptyset. \tag{4.5}$$

The concentration of the complex $S_i\sim$CFSE at time t is denoted by $X_{i,t}$. The measured quantity is the sum of CFSE induced fluorescence, which is proportional to the overall CFSE concentration, [$\bullet\sim$CFSE], in a cell,

$$X_t = \sum_i^{d_S} X_{i,t}. \tag{4.6}$$

Despite the fact that the reaction events are stochastic in nature and can be described by a Markov jump process, as the number of CFSE molecules is large, it is plausible to consider the expected values of $\mathbb{E}[X_t]$ and $\mathbb{E}[X_{i,t}]$.

As the individual degradation processes are linear Markov jump processes, the dynamics of the expected values are governed by

$$\frac{d\mathbb{E}[X_{i,t}]}{dt} = -k_i\mathbb{E}[X_{i,t}], \quad i = 1, \dots, d_S \tag{4.7}$$

(see, e.g., (Hespanha, 2007)), yielding $\mathbb{E}[X_{i,t}] = e^{-k_i t}\mathbb{E}[X_{i,0}] = e^{-k_i t}P(S_i)\mathbb{E}[X_0]$. On the other hand, the expected value $\mathbb{E}[X_t]$ is the sum of the expected values $\mathbb{E}[X_{i,t}]$. Therefore, the time dependent expectation is X_t is

$$\mathbb{E}[X_t] = \sum_{i=1}^{d_S} \mathbb{E}[X_{i,t}] = \left(\sum_i^{d_S} e^{-k_i t}P(S_i)\right)\mathbb{E}[X_0]. \tag{4.8}$$

This already shows, that the overall degradation process of

$$\bullet\sim\text{CFSE} \xrightarrow{k(t)} \emptyset \tag{4.9}$$

can in general not be described by a linear degradation process,

$$\frac{d\mathbb{E}[X_t]}{dt} = -k(t)\mathbb{E}[X_t], \tag{4.10}$$

with constant degradation rate, $k(t) = k$. A constant degradation rate would imply that $\forall t > 0 : \sum_i^{d_S} e^{-k_i t}P(S_i) = e^{-kt}$. It can be shown easily that this equation only holds if CFSE solely binds to molecules with identical degradation rates, $k_i = k$ for $P(S_i) \neq 0$. As the CFSE binding is known to be highly unspecific, this will not be the case.

Given this first insight, which proves that most models so far relied on an incorrect assumption concerning the degradation rate, it would be important to (1) study whether the overall degradation of CFSE is a linear process (4.10), and (2) what the time dependent rate is as a function of $P(S_i)$ and k_i. To approach these questions, note that the time derivative of $\mathbb{E}[X_t]$ is the sum of the time derivatives of the individual species,

$$\frac{d\mathbb{E}[X_t]}{dt} = \sum_{i=1}^{d_S} \frac{d\mathbb{E}[X_{i,t}]}{dt} = \sum_{i=1}^{d_S} -k_i\mathbb{E}[X_{i,t}]. \tag{4.11}$$

By plugging the analytical solution of the degradation processes of the individual species into this equation we obtain

$$\frac{d\mathbb{E}[X_t]}{dt} = \left(\sum_{i=1}^{d_S} -k_i e^{-k_i t} P(S_i)\right) \mathbb{E}[X_0]\,. \tag{4.12}$$

This provides us with a second equation for the time evolution of $\mathbb{E}[X_t]$. By solving (4.8) for $\mathbb{E}[X_0]$ and inserting it in (4.14), we obtain

$$\frac{d\mathbb{E}[X_t]}{dt} = \left(\frac{\sum_{i=1}^{d_S} k_i e^{-k_i t} P(S_i)}{\sum_{i=1}^{d_S} e^{-k_i t} P(S_i)}\right) \mathbb{E}[X_t]\,. \tag{4.13}$$

This verifies that the overall CFSE degradation process is indeed linear, but possesses the time-dependent degradation rate

$$k(t) = \frac{\sum_{i=1}^{d_S} k_i e^{-k_i t} P(S_i)}{\sum_{i=1}^{d_S} e^{-k_i t} P(S_i)}. \tag{4.14}$$

The degradation rate $k(t)$ depends nonlinearly on the distribution of degradation rates k_i. Unfortunately, for non-trivial distributions of k_i no compact solution of (4.14) could be found. However, an analysis of $k(t)$ reveals that for short times $t \ll 1$, $k(t)$ will be dominated by the degradation rates of species with short lifespans ($k_i \gg 1$). In contrast, at late times $t \gg 1$, the degradation rate is predominantly influenced by the degradation rates of long-living proteins ($k_i \ll 0$). Hence, the model explains the rapid decline of CFSE induced fluorescence directly after labeling, as well as the detectable levels after weeks (Banks *et al.*, 2011; Lyons & Parish, 1994). In addition, the model provides a mathematical justification for modeling CFSE degradation as a Gompertz decay process, $k(t) = k_{\max} e^{-k_T t}$, as proposed in (Banks *et al.*, 2011). Gompertz decay processes are capable of describing the short- and long-term dynamics of $k(t)$ (4.14).

Beyond CFSE degradation, some studies also consider incomplete binding of CFSE to intracellular proteins, as well as secretion of free CFSE and CFSE-protein complexes. This can easily be incorporated into the above model, if the secretion process is assumed to be linear. In this case we obtain equations of type (4.12) and (4.14), which additionally contain the probability of different secretion rates. Following the argumentation in (Banks *et al.*, 2011) that the process of CFSE secretion is faster than protein degradation, we find that secretion could explain the rapid decay during the first hours, which is also discussed by Banks *et al.* (2011).

The derived mathematical model of CFSE degradation does not only support existing hypotheses, but can also be used the infer information about cellular degradation. Assuming that CFSE binding to intracellular proteins is unspecific, the observed time dependence of $k(t)$ enables the reconstruction of the degradation rate distribution. This distribution is of interest in many fields, e.g., cancer cell biology (Eden *et al.*, 2011), as changes in the protein degradation indicate global alteration in cell properties. In contrast to the time consuming assessment of individual half-life times performed by (Eden *et al.*, 2011), the observation of the time-dependent CFSE degradation might allow for a global assessment of degradation processes.

4.2.2 Modeling division- and label-structured populations

The single cell dynamics studied earlier are now employed to describe the proliferation dynamics of CFSE labeled cell populations. This requires the consideration of two important distinct features on the population level:

- the label concentration x and
- the number of cell divisions i a cell has undergone.

The importance of the label concentration $x \in \mathbb{R}_+$ arises from the fact that this is the quantity which can be observed, e.g., using flow cytometry (Hawkins *et al.*, 2007). On the other hand, a direct observation of the number of cell divisions $i \in \mathbb{N}_0$ a cell has undergone is in general not possible, though the division number often plays a crucial role within the model. A cell which has divided once is expected to have different properties, e.g., a different division rate, than a cell which has already divided several dozen times (Hayflick, 1965; Kassem *et al.*, 1997). This induces discrete cell-to-cell variability on the otherwise homogeneous population.

In this thesis, we propose a model which captures both features of cells, distinct division numbers as well as distinct label concentrations among cells. Therefore, instead of a single PDE model describing the label dynamics of the overall population, a PDE model is defined for every subpopulation. Thereby, the i-th subpopulation contains the cells which have divided i times. Cell division generates a flux from subpopulation i to subpopulation $i+1$, thus inducing coupling. The system of coupled PDEs is given by

$$\begin{aligned} i = 0: &\quad \frac{\partial n(x,0|t)}{\partial t} + \frac{\partial (v(t,x) n(x,0|t))}{\partial x} = -\left(\alpha_0(t) + \beta_0(t)\right) n(x,0|t) \\ \forall i \geq 1: &\quad \frac{\partial n(x,i|t)}{\partial t} + \frac{\partial (v(t,x) n(x,i|t))}{\partial x} = -\left(\alpha_i(t) + \beta_i(t)\right) n(x,i|t) + 2\gamma\alpha_{i-1}(t) n(\gamma x, i-1|t), \end{aligned} \tag{4.15}$$

with initial conditions

$$i = 0 : n(x,0|0) \equiv n_0(x), \quad \forall i \geq 1 : n(x,i|0) \equiv 0.$$

In this system, $n(x,i|t)$ denotes the label density in the i-th subpopulation at time t. This label density is described by a number density function, meaning that $\int_{x\in\Omega} n(x,i|t)dx$ provides the number of cells with $x \in \Omega$ and i divisions at time t.

The structure of the models for the individual subpopulations is highly similar to a single PDE which is employed in label-structured models (Luzyanina *et al.*, 2007b). The fluxes influencing the label distribution $n(x|i,t)$ are:

- $\partial(v(t,x)n(x,i|t))/\partial x$, decay of label x in each cell with label loss rate $v(t,x)$.
- $-\left(\alpha_i(t) + \beta_i(t)\right) n(x,i|t)$, disappearance of cells from the i-th subpopulation due to cell division with rate $\alpha_i(t)$ and due to cell death with rate $\beta_i(t)$.
- $2\gamma\alpha_{i-1}(t)n(\gamma x, i-1|t)$, appearance of two cells due to cell division in the $(i-1)$th subpopulation with division rate $\alpha_{i-1}(t)$. The factor $\gamma \in (1,2]$ is the rate of label dilution due to cell division (cf. (Banks *et al.*, 2010; Luzyanina *et al.*, 2009)).

It has to be emphasized that the division rates $\alpha_i(t) : \mathbb{R}_+ \rightarrow \mathbb{R}_+$, as well as the death rates $\beta_i(t) : \mathbb{R}_+ \rightarrow \mathbb{R}_+$ may depend on the division number i and the time t. To ensure existence and uniqueness of the solutions we require $\alpha_i(t), \beta_i(t) \in C^1$. As it is assumed that the labeling does not affect the cell functions, we do not allow α_i and β_i to depend on the label concentration x. Furthermore, due to the analysis performed in the previous section, only label loss rates are considered which describe a linear degradation

$$\nu(x) = -k(t)x. \tag{4.16}$$

In particular, the cases of constant degradation processes (Banks *et al.*, 2010; Luzyanina *et al.*, 2009, 2007b), $k(t) = k$ (const.), and Gompertz decay processes, $k(t) = k_{\max} e^{-k_T t}$ (Banks *et al.*, 2011), will be discussed.

Note that by construction, model (4.15) provides information about cell numbers and label distribution for the overall as well as for individual subpopulations. Hence, it combines advantages of common ODE models (De Boer *et al.*, 2006; Deenick *et al.*, 2003; Luzyanina *et al.*, 2007a; Revy *et al.*, 2001) and common PDE models (Banks *et al.*, 2011, 2010; Luzyanina *et al.*, 2009, 2007b) of cell populations and permits more biologically plausible degrees of freedom than either of them. In detail, the available information is:

Number of cells in the subpopulations: Given $n(x, i|t)$, the number of cells which divided i times until time t is

$$N(i|t) = \int_{\mathbb{R}_+} n(x, i|t) dx. \tag{4.17}$$

This number is required to understand the composition of the overall population, which is in turn necessary to analyze the proliferation potential.

Normalized label density in the subpopulations: Given $n(x, i|t)$, the probability density of different label concentrations within the i-th subpopulation can be computed. For all pairs $(i, t) \in \mathbb{N}_0 \times \mathbb{R}_+$ for which $N(i|t) > 0$, the probability of certain label concentration, x is

$$p(x|i, t) = \frac{n(x, i|t)}{N(i|t)}. \tag{4.18}$$

This probability density, $p(x|i, t)$, is the probability density that a cell which is in the i-th subpopulation has a label concentration x. Thus, given a cell in the i-th subpopulation, its probability of having label concentration $\tilde{x} \in [x, x + \Delta x]$ is

$$\mathrm{Prob}(\tilde{x} \in [x, x + \Delta x]) = \int_x^{x+\Delta x} p(\tilde{x}|i, t) d\tilde{x}. \tag{4.19}$$

Besides the properties of the subpopulations, the model permits also the analysis of the properties of the overall population. The unnormalized label density in the overall cell population $n(x|t)$ is given by

$$n(x|t) = \sum_{i \in \mathbb{N}_0} n(x, i|t). \tag{4.20}$$

An autofluorescence corrupted version of this distribution can be assessed using CFSE or BrdU based proliferation assays. In addition, the overall population size

$$N(t) = \int_0^{\infty} n(x|t) dx = \sum_{i \in \mathbb{N}_0} N(i|t) \tag{4.21}$$

can be determined experimentally using simple cell counting. As there is currently no direct cell division marker available, experimental assessment of the subpopulation sizes or of the label distribution within the subpopulations is in general not feasible. All common experimental techniques only provide data marginalized over the division number i.

Remark 4.1. *We note that the cell numbers, $N(t)$ and $N(i|t)$, are real and not integer-valued. Hence, our model does, like previous ODE and PDE-based models, not capture the discreteness of the cell numbers. However, this discrepancy is negligible if the number of cells is large enough.*

4.2.3 Analysis of the division- and label-structured population model

Besides the advantages the DLSP model offers, its potential drawback is its complexity. The model is a system of coupled PDEs, which are in general difficult to analyze, and their simulation is often computationally demanding or even intractable. In the following, it is shown that these problems can be resolved for the DLSP model (4.15). The approach presented allows to efficiently compute the solution of the DLSP model without solving a system of coupled PDEs.

Solution of the DLSP model via decomposition

In order to provide an efficient method for computing the solution of (4.15), we define the initial number of cells

$$N_0 = \int_{\mathbb{R}_+} n_0(x)dx \tag{4.22}$$

and the initial label density

$$p_0(x) = \begin{cases} \dfrac{n_0(x)}{N_0} & \text{for } N_0 > 0 \\ 0 & \text{otherwise,} \end{cases} \tag{4.23}$$

according to (4.17) and (4.18). Given these definitions the following theorem holds:

Theorem 4.2. *The solution of model* (4.15) *is*

$$\forall i: \quad n(x,i|t) = N(i|t)p(x|i,t) \tag{4.24}$$

in which:

(i) $N(i|t)$ *is the solution of the system of ODEs:*

$$\begin{aligned} i = 0: \quad & \frac{dN}{dt}(0|t) = -\left(\alpha_0(t) + \beta_0(t)\right) N(0|t), \\ \forall i \geq 1: \quad & \frac{dN}{dt}(i|t) = -\left(\alpha_i(t) + \beta_i(t)\right) N(i|t) + 2\alpha_{i-1}(t)N(i-1|t) \end{aligned} \tag{4.25}$$

with initial conditions: $N(0|0) = N_0$ *and* $\forall i \geq 1: N(i|0) = 0$.

(ii) $p(x|i,t)$ *is the solution of the PDE:*

$$\forall i: \quad \frac{\partial p(x|i,t)}{\partial t} - k(t)\frac{\partial (xp(x|i,t))}{\partial x} = 0 \tag{4.26}$$

with initial conditions $\forall i: p(x|0,t) \equiv \gamma^i p_0(\gamma^i x)$.

The state variables $N(i|t)$ and $p(x|i,t)$ of the ODE system and the PDEs correspond to the number of cells (4.17) and the normalized label density (4.18) in the i-th subpopulation, respectively.

Proof. To prove that Theorem 4.2 holds, (4.24) - (4.26) are inserted in (4.15) and it is shown that the resulting equation holds. The proof is only shown for $i \geq 1$, since the case $i = 0$ can be treated analogously.

Inserting (4.24) in (4.15) for $i \geq 1$ yields

$$\begin{aligned}\frac{\partial(N(i|t)p(x|i,t))}{\partial t} - k\frac{\partial(xN(i|t)p(x|i,t))}{\partial x} = \\ -(\alpha_i(t)+\beta_i(t))\,N(i|t)p(x|i,t) + 2\gamma\alpha_{i-1}(t)N(i-1|t)p(\gamma x|i-1,t).\end{aligned} \tag{4.27}$$

The left hand side of this equation can be reformulated:

$$\begin{aligned}&\frac{\partial(N(i|t)p(x|i,t))}{\partial t} - k\frac{\partial(xN(i|t)p(x|i,t))}{\partial x} \\ &= \frac{dN(i|t)}{dt}p(x|i,t) + N(i|t)\frac{\partial n_i}{\partial t} - kN(i|t)\frac{\partial(xp(x|i,t))}{\partial x} \\ &= \frac{dN(i|t)}{dt}p(x|i,t) + N(i|t)\left(\frac{\partial n_i}{\partial t} - k\frac{\partial(xp(x|i,t))}{\partial x}\right) \\ &\overset{(4.26)}{=} \frac{dN(i|t)}{dt}p(x|i,t).\end{aligned} \tag{4.28}$$

By inserting this result in (4.27) and substituting $dN(i|t)/dt$ with (4.25), we obtain

$$\begin{aligned}(-(\alpha_i+\beta_i)\,N(i|t) + 2\alpha_{i-1}N(i-1|t))\,p(x|i,t) = \\ -(\alpha_i+\beta_i)\,N(i|t)p(x|i,t) + 2\gamma\alpha_{i-1}N(i-1|t)p(\gamma x|i-1,t),\end{aligned} \tag{4.29}$$

which can be simplified to

$$p(x|i,t) = \gamma p(\gamma x|i-1,t). \tag{4.30}$$

Using the analytical expression for $p(x|i,t)$ derived in Appendix A, it can be proven easily that (4.30) holds, which concludes the proof of Theorem 4.2. □

Note that it can also be verified that (4.30) holds if and only if the label loss rate $\nu(t,x)$ is linear in x, $\nu(t,x) = -k(t)x$. Furthermore, we would like to emphasize that according to (4.20), $p(x|i,t)$ is not specified for pairs (i,t) for which $n(x,i|t) \equiv 0$, e.g., for any $i \geq 1$ and $t = 0$. This is the case as for such pairs (i,t), no cells exist in the respective subpopulation i, implying that $N(i|t) = 0$. Hence, (4.24) simplifies to $0 = 0 \cdot p(x|i,t)$, allowing any specification for $p(x|i,t)$, among others our choice, e.g., for any $i \geq 1$ and $t = 0$: $p(x|i,0) \equiv \gamma^i p_0(\gamma^i x)$. This choice is consistent with the state of the biological system, as the normalized label distribution is undefined.

With Theorem 4.2, the original system of coupled PDEs can be decomposed into a system of ODEs (4.25) and a set of decoupled PDEs (4.26). This means that the size of the individual subpopulations can be decoupled from the label dynamics. This already tremendously simplifies the analysis, but a further simplification is possible:

Corollary 4.3. *The solution of model* (4.15) *is*

$$\forall i: \quad n(x,i|t) = N(i|t)\gamma^i e^{\int_0^t k(\tau)d\tau} p_0(\gamma^i e^{\int_0^t k(\tau)d\tau} x), \tag{4.31}$$

in which $N(i|t)$ is the solution of the ODE (4.25).

Proof. To prove Corollary 4.3 note that the PDE (4.26) is linear. Thus, the method of characteristics (Evans, 1998) can be employed to obtain an analytical solution (Appendix A). This yields

$$\forall i : p(x|i,t) = \gamma^i e^{\int_0^t k(\tau)d\tau} n_0(\gamma^i e^{\int_0^t k(\tau)d\tau} x), \tag{4.32}$$

which can be inserted into (4.24), proving Corollary 4.3. □

The general solution $n(x, i|t)$ simplifies in cases of specific choices for $k(t)$. A constant degradation rate yields

$$\forall i : p(x|i,t) = \gamma^i e^{kt} p_0(\gamma^i e^{kt} x), \tag{4.33}$$

while for a Gompertz decay process one obtains,

$$\forall i : p(x|i,t) = \gamma^i e^{\frac{k_{\max}}{k_T}(1-e^{-k_T t})} p_0(\gamma^i e^{\frac{k_{\max}}{k_T}(1-e^{-k_T t})} x). \tag{4.34}$$

Corollary 4.3 provides a solution for any label degradation rates, including those considered in (Banks *et al.*, 2012; Thompson, 2012).

By solving the decoupled PDEs analytically, the solution of the DLSP model can be obtained in terms of the solution of a system of ODEs. This reduces the complexity drastically and enables also a compact representation of the overall label density $n(x|t)$:

Corollary 4.4. *The overall label density* (4.20) *is*

$$n(x|t) = \sum_{i\in\mathbb{N}_0} N(i|t) p(x|i,t) = \sum_{i\in\mathbb{N}_0} N(i|t) \gamma^i e^{\int_0^t k(\tau)d\tau} p_0(\gamma^i e^{\int_0^t k(\tau)d\tau} x), \tag{4.35}$$

in which $N(i|t)$ is the solution of the ODE (4.25).

Proof. By substituting (4.31) into (4.20) we obtain (4.35), which proves Corollary 4.4. □

Given Corollary 4.3 and 4.4, it is apparent that merely the ODE system (4.25) has to be solved in order to compute the solution of the DLSP model. This problem is approached in the remainder of this section.

Calculation of the subpopulation sizes

To solve the ODE system (4.25), note that the number of cell in subpopulation i merely depends on the number of cells in subpopulation $i - 1$. This chain-like structure enables the solution of $N(i|t)$ via recursion. By doing so, analytical solutions for the ODE system have been found for two cases (Luzyanina *et al.*, 2007a; Revy *et al.*, 2001):

Lemma 4.5. *Given that* $\forall i \in \mathbb{N}_0 : \alpha_i(t) = \alpha \geq 0 \wedge \beta_i(t) = \beta > 0$, *the solution of* (4.25) *is:*

$$N(i|t) = \frac{(2\alpha t)^i}{i!} e^{-(\alpha+\beta)t} N_0. \tag{4.36}$$

This result has been derived in (Revy *et al.*, 2001), where the authors studied this ODE system to model the number of cells that have undergone a certain number of divisions, without modeling label dynamics. For later use in this thesis, a generalized derivation is provided in Appendix B.

Lemma 4.6. *Given that* $\forall i \in \mathbb{N}_0 : \alpha_i(t) = \alpha_i \geq 0 \ \wedge \ \beta_i(t) = \beta_i > 0$ *and* $\forall i, j \in \mathbb{N}_0, i \neq j : \alpha_i + \beta_i \neq \alpha_j + \beta_j$, *then the solution of* (4.25) *is:*

$$\begin{aligned} i = 0 &: N(0|t) = e^{-(\alpha_0+\beta_0)t} N_0 \\ \forall i \geq 1 &: N(i|t) = 2^i \left(\prod_{j=1}^{i} \alpha_{j-1} \right) D_i(t) N_0 \end{aligned} \tag{4.37}$$

in which

$$D_i(t) = \sum_{j=0}^{i} \left[\left(\prod_{\substack{k=0 \\ k \neq j}}^{i} ((\alpha_k + \beta_k) - (\alpha_j + \beta_j)) \right)^{-1} e^{-(\alpha_j+\beta_j)t} \right].$$

Solution (4.37) was first stated in (Luzyanina *et al.*, 2007a) and for completeness the proof is provided in Appendix C. It basically employs mathematical induction in the frequency domain. Despite the prerequisites, this result is quite powerful as for almost all cases of time invariant division number-dependent parameters α_i and β_i the ODE system (4.25) can be solved analytically.

In cases in which neither conditions for Lemma 4.5 nor 4.6 are met, the solution of (4.25) can be computed using numerical integration. This is possible if only the sizes of the first S subpopulations, $N(0|t)$, $N(1|t)$, ..., $N(S|t)$, are of interest, where S is finite.

Truncation of division numbers in the population model

In Section 4.2.3 a decomposition approach has been described to decouple the size of the subpopulations from the label distribution in the individual subpopulations. While this simplifies the computation of the properties of individual subpopulations drastically, the analysis of the overall label density and of the overall population size still requires the calculation of an infinite sum (4.35). Even in cases for which the individual subpopulation sizes are available analytically (see (4.36) and (4.37)), we could not derive a closed form solution for $n(x|t)$. Therefore, in this section we present a method to find an approximation of $n(x|t)$ of the form

$$\hat{n}_S(x|t) = \sum_{i=0}^{S-1} n(x, i|t) = \sum_{i=0}^{S-1} N(i|t) p(x|i, t) \tag{4.38}$$

with truncation index $S \geq 0$. Instead of considering an infinite number of subpopulations, only the first S subpopulations are taken into account. While it might be argued that a bound S can be determined from experimental data collected in proliferation assays (Banks *et al.*, 2012; Thompson, 2012), this is not true for long times. In case of long observation intervals, the autofluorescence – which will be discussed in Section 4.2.5 – avoids an estimation of S. Thus, reliable selection rules for the truncation index S are necessary.

In order to approximate $n(x|t)$ with arbitrary precision by the truncated sum $\hat{n}_S(x|t)$, convergence of (4.20) and (4.21) with respect to the subpopulation index i is required and can be proven:

Theorem 4.7. *The sums* (4.20) *converge for any finite time* T, *if there exist*

$$\begin{aligned} \alpha_{\sup} &= \sup_{t\in[0,T], i\in\mathbb{N}_0} \alpha_i(t) \geq 0, \\ \alpha_{\inf} &= \inf_{t\in[0,T], i\in\mathbb{N}_0} \alpha_i(t) \geq 0, \\ \beta_{\inf} &= \inf_{t\in[0,T], i\in\mathbb{N}_0} \beta_i(t) > 0. \end{aligned} \tag{4.39}$$

The proof of Theorem 4.7 is provided in Appendix E. It employs a system of ODEs whose states are an upper bound for the states of (4.25), and which can be solved analytically. Given these upper bounds the comparison theorem for series (Knopp, 1964) can be used to verify convergence. Note that Theorem 4.7 is powerful as it holds for all biologically plausible functions $\alpha_i(t)$ and $\beta_i(t)$.

Given convergence the question arises how large the truncation index S must be to ensure a predefined error bound at a given time.

Theorem 4.8. *Given a truncation index S and a time T, as well as α_{inf}, α_{sup}, and β_{inf} as defined in Theorem 4.7, the truncation error is upper bounded by $E_S(T)$:*

$$\frac{\|n(x|T) - \hat{n}_S(x|T)\|_1}{\|n(x|0)\|_1} \leq E_S(T) = \left(e^{2\alpha_{\text{sup}}T} - \sum_{i=0}^{S-1} \frac{(2\alpha_{\text{sup}}T)^i}{i!} \right) e^{-(\alpha_{\text{inf}}+\beta_{\text{inf}})T}. \tag{4.40}$$

To prove Theorem 4.8, we show that $\|n(x|t) - \hat{n}_S(x|t)\|_1 = \sum_{i=S+1}^{\infty} N(i|t)$. This sum can be upper bounded for all biologically plausible functions $\alpha_i(t)$ and $\beta_i(t)$ using the ODE system employed to verify Theorem 4.7. The full proof is provided in Appendix F. Note that the bound (4.40) is tight as equality holds for: $\forall i \in \mathbb{N}_0 : \alpha_i(t) = \alpha \ \wedge \ \beta_i(t) = \beta$.

Remark 4.9. *In this work we considered an error bound which is relative to the initial condition. This is reasonable as for this system the superposition principle holds and the relative truncation error $\frac{\|n(x|T)-\hat{n}_S(x|T)\|_1}{\|n(x|0)\|_1}$ is thus independent of $\|n(x|0)\|_1$.*

Given Theorem 4.8, an upper bound S can be derived which ensures that a relative error is bounded by ϵ:

Corollary 4.10. *Assuming that α_{inf}, α_{sup}, and β_{inf} exist as defined in Theorem 4.7, the error bound*

$$\frac{\|n(x|T) - \hat{n}_S(x|T)\|_1}{\|n(x|0)\|_1} \leq \epsilon \tag{4.41}$$

holds if

$$\left(e^{2\alpha_{\text{sup}}T} - \sum_{i=0}^{S-1} \frac{(2\alpha_{\text{sup}}T)^i}{i!} \right) e^{-(\alpha_{\text{inf}}+\beta_{\text{inf}})T} \leq \epsilon. \tag{4.42}$$

Proof. Corollary 4.10 follows directly from Theorem 4.8, using $\frac{\|n(x|T)-\hat{n}_S(x|T)\|_1}{\|n(x|0)\|_1} \leq E_S(T) \leq \epsilon$. □

Despite the generality of Theorem 4.8 and Corollary 4.10, there is a minor disadvantage. For the considered system class, no explicit expression for S has been found. Rather, the minimum truncation index S which is required to ensure a certain error bound has to be found iteratively by increasing or decreasing S based on the current error. Fortunately, this search is computationally cheap as it is not necessary to solve a system of ODEs or PDEs, but the error bound is available analytically.

A study of the a priori error bound (4.42) shows that if the acceptable relative error ϵ is kept constant, the truncation index S grows monotonically as a function of the final simulation time, T. This is due to the exponential growth of $e^{2\alpha_{\text{sup}}T}$ vs. the polynomial growth of $\sum_{i=0}^{S-1} \frac{(2\alpha_{\text{sup}}T)^i}{i!}$. Merely for cases in which $2\alpha_{\text{sup}} \leq \alpha_{\text{inf}} + \beta_{\text{inf}}$, S does not have to increase arbitrarily over time but stays bounded, as under these conditions the population dies out. Note that the increase of S is often not critical. Due to label dilution in general only the first six or eight cell divisions can be observed (Hawkins *et al.*, 2007), which limits the timespan of interest and therefore the required truncation index S. Simulation studies showed that for a realistic setup the simulation of the first 20 subpopulations is completely sufficient (Hasenauer *et al.*, 2012c).

Aside from an approximation of the population density $n(x|t)$, also approximations for the overall population size $N(t)$ may be necessary to compare model predictions to measurements. A reasonable choice is

$$\hat{N}_S(t) = \int_{\mathbb{R}_+} \hat{n}_S(x|t)dx = \sum_{i=1}^{S} N(i|t), \tag{4.43}$$

which possesses the same convergence property and error bound as $\hat{n}_S(x|t)$.

4.2.4 Comparison of different proliferation models

In the last section we have analyzed the DLSP model and outlined a method to solve it. The question which remains open is how the DLSP model and its solution relate to existing population models for cell proliferation. To answer this question we confine ourselves to the in our opinion most common models, the exponential growth (EG) model , the division-structured population (DSP) model and the label-structured population (LSP) model:

- EG model: A single ODE describing the dynamics of the overall population size (Zwietering *et al.*, 1990).
- DSP model: A system of ODEs describing the dynamics of the number of cells contained in the individual subpopulations. Similar to the DLSP model, each subpopulation is defined via a common number of cell divisions (De Boer *et al.*, 2006; Revy *et al.*, 2001).
- LSP model: A PDE describing the dynamics of the label density in the overall population (Banks *et al.*, 2011, 2010; Luzyanina *et al.*, 2009, 2007b).

These models are used in many more publications than cited here and various extensions of these models exist.

Relation between the EG model and the DLSP model

The EG model is the simplest available model which describes population dynamics. It has only one state variable, which corresponds to the size of the overall cell population. In general, the EG model is written as

$$\frac{dN^{\mathrm{EG}}(t)}{dt} = \phi(t)N^{\mathrm{EG}}(t), \quad N^{\mathrm{EG}}(0) = N_0^{\mathrm{EG}}, \tag{4.44}$$

in which $\phi(t)$ is the effective growth rate. A common choice is $\phi(t) = e^{\phi_1 - \phi_2 t}$ which results in a Gompertz equation (Zwietering *et al.*, 1990).

As the EG model only describes the overall population size, it is contained in the DLSP model. By choosing, e.g., $\alpha_i(t) = \phi(t)$, $\beta_i(t) = 0$ and $N_0 = N_0^{\mathrm{EG}}$, the overall population size

$N(t)$ predicted by the DLSP model is equivalent to $N^{EG}(t)$. This can be shown using the time derivative of $N(t)$,

$$\frac{dN(t)}{dt} = \sum_{i \in \mathbb{N}_0} \frac{dN(i|t)}{dt} = \phi(t) \sum_{i \in \mathbb{N}_0} N(i|t) = \phi(t)N(t), \tag{4.45}$$

which has the initial condition $N(0) = \sum_{i \in \mathbb{N}_0} N(i|0) = N_0 = N_0^{EG}$.

Relation between the DSP model and the DLSP model

In contrast to the EG model, the DSP model resolves the subpopulations, and the state variables $N^{DSP}(i|t)$ correspond to the number of cells which have divided i times. To our knowledge this model has first been proposed in (Revy *et al.*, 2001) and its most common form is equal to (4.25). Thus, the DSP model is contained in the DLSP model and is obtained by marginalization over the label concentration x. Actually, according to Theorem 4.2, a DSP model is solved to compute the solution of the DLSP model. As for the PDE component of the DLSP model an analytical expression can be derived (Corollary 4.3), solving the DLSP model poses basically the same numerical challenge as solving the DSP model.

Relation between the LSP model and the DLSP model

For the comparison of model predictions and labeling experiments with CFSE or BrdU, the LSP model has been introduced (Banks *et al.*, 2011, 2010; Luzyanina *et al.*, 2009, 2007b). The state variable of the LSP model denote the label density $M^{LSP}(t, x)$ in the population. In general, the evolution of $M^{LSP}(t, x)$ is modeled by the PDE

$$\frac{\partial n^{LSP}(x|t)}{\partial t} + \frac{\partial(\nu(x)n^{LSP}(x|t))}{\partial x} = -\left(\alpha(t, x) + \beta(t, x)\right) n^{LSP}(x|t) + 2\gamma\alpha(t, x)n^{LSP}(\gamma x|t), \tag{4.46}$$

with initial condition $n^{LSP}(x|0) \equiv n_0^{LSP}(x|t)$. As this model allows for label dependent division and death rates, $\alpha(t, x)$ and $\beta(t, x)$, it is in this respect more general than the DLSP model.

However, it is not obvious why the cell division or death rates should depend on the label concentration. If the experiments are performed at low label concentrations far from the toxic regime, the population dynamics should be independent of the labeling (Lyons & Parish, 1994; Matera *et al.*, 2004). In particular, complex dependencies of $\alpha(t, x)$ and $\beta(t, x)$ on the label concentrations x, like those shown in (Banks *et al.*, 2010), are hard to argue. Additionally, a recent study supports that the introduced nonlinearities are correlated with the division number (Banks *et al.*, 2010).

Therefore, we just consider division and death rates which solely depend on time t, $\alpha(t)$ and $\beta(t)$. As proven in Appendix G, for this case, the solution $n(x|t)$ of the DLSP model, with $\alpha_i(t) = \alpha(t)$ and $\beta_i(t) = \beta(t)$ and $n_0(x) \equiv n_0^{LSP}(x)$, is equivalent to $n^{LSP}(x|t)$. This shows that under these assumptions, the information provided by the LSP model is a subset of the information available from the DLSP model. This renders the DLSP model more useful, as also subpopulation sizes are accessible.

Furthermore, for time dependent $\alpha(t)$ and $\beta(t)$, the solution of the DLSP model can be approximated by a low-dimensional ODE system (Theorems 4.7 and 4.8). Hence, instead of computing $n^{LSP}(x|t)$ using a PDE solver as done in all available publications, one may solve only a low-dimensional ODE system. Using the analytical results for the ODE system (4.25) even analytical solutions are available, e.g.,

$$n^{\mathrm{LSP}}(x|t) = e^{-(\alpha+\beta)t} e^{kt} \left(\sum_{i \in \mathbb{N}_0} \frac{(2\alpha\gamma t)^i}{i!} n_0^{\mathrm{LSP}}(\gamma^i e^{kt} x) \right), \tag{4.47}$$

for constant rates α and β. Although this result for the LSP model may be helpful to study various systems, we have not found it in the literature yet. The reason might be that a direct derivation of (4.47) is rather complex, whereas the study of the DLSP model renders it straightforward.

Clearly, label dependent cell division and death rates or constant label loss rates were not considered here, in contrast to what was done in (Banks *et al.*, 2010; Luzyanina *et al.*, 2009, 2007b). This was avoided as the decomposition of the solution shown in Section 4.2.3 becomes impossible and solving the DLSP model gets computationally challenging. Nevertheless, the loss of these degrees of freedom is compensated by allowing for biologically more plausible division dependent cell parameters in the DLSP model.

DLSP model as a unifying modeling framework

The implications of the findings above are that the three most prevalent classes of population models are captured by the DLSP model. Furthermore, the DLSP model is more general, as label distributions and division dependent parameters may be considered, which are both important and well motivated from a biological point of view. Figure 4.5 illustrates these relations and shows how the EG model, the DSP model, and the LSP model may be constructed from the DLSP model via marginalization.

In contrast to the generality, the simulation effort increases only marginally when studying the DLSP model instead of the DSP model or the LSP model. This is due to the decomposition into a system of ODEs (which is equivalent to the DSP model), and a single set of PDEs. The set of PDEs can be solved analytically, and in several cases even analytical solutions for the ODE exist, facilitating an analytical solution of the overall system. Such analytical solutions can then be used to determine previously unknown analytical solutions for the LSP model, e.g., like (4.47).

Remark 4.11. *Obviously, there exist extensions of the LSP model and the DSP model which are not captured by the current version of the DLSP model. Examples are the aforementioned label concentration dependent division and death rates for the LSP model (Banks* et al., *2010; Luzyanina* et al., *2009, 2007b) as well as DSP models with recruitment delay (De Boer* et al., *2006; León* et al., *2004). While the DLSP model can easily be extended to take such effects into account, the numerical analysis will become more challenging.*

4.2.5 Prediction of the measured fluorescence distribution

In the previous sections, we established the DLSP model, describing the dynamics of the label distribution, and related it to existing models. Next, we will establish the relation to the measured data.

As outlined in Section 4.1.2, to obtain quantitative information about the proliferation dynamics, the fluorescent levels of individual cells are assessed using flow cytometry (Hawkins *et al.*, 2007). The fluorescence level of an individual cell summarizes the CFSE induced

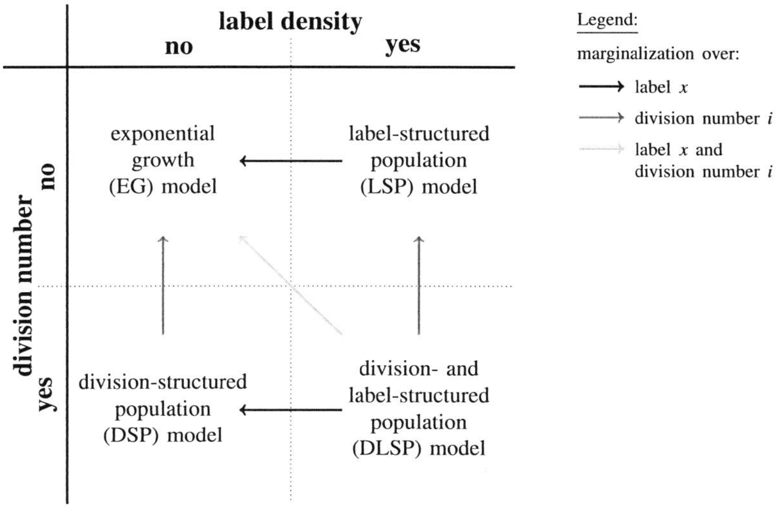

Figure 4.5: Illustration of the relation between the exponential growth model, the division-structured population model, the label-structured population model, and the division- and label structured population model. The models are distinguished using two properties, the availability of division numbers (vertical axis) and of information about the label distribution (horizontal axis). Arrows indicate whether and arrow labels describe how a model can be obtained from another model. It is apparent that the DLSP model is the most general model, as all remaining models can be constructed from it via marginalization.

fluorescence and the autofluorescence, $\bar{Y} = Y + Y_a$ (4.1). While the CFSE induced fluorescence is completely determined by the CFSE concentration, $Y = cX$, the autofluorescence is a stochastic variable, $Y_a \sim p_a(y_a)$.

Relation of the CFSE concentration and the CFSE induced fluorescence

In most studies, neither the CFSE concentration, X, nor the proportionality constant, c, are of interest. Merely, the information about the CFSE induced fluorescence, $Y = cX$, and its distribution within the cell population shall be used to infer the proliferation properties. This can be employed to avoid the estimation of c.

The distribution of the CFSE induced fluorescence, $n(y, i|t)$, is directly related to $n(x, i|t)$ via

$$\forall t \in \mathbb{R}_+, i \in \mathbb{N}_0 : \qquad n(x, i|t) = \frac{1}{c} n(y = cx, i|t), \tag{4.48}$$

which follows directly from (4.1). Employing this, it can be shown that the PDE model (4.15) also governs the dynamics of $n(y, i|t)$, when substituting x by y and $k(t)$ by $\tilde{k}(t) = ck(t)$:

$$i = 0: \quad \frac{\partial n(y,0|t)}{\partial t} - \tilde{k}(t)\frac{\partial(yn(y,0|t))}{\partial y} = -(\alpha_0(t) + \beta_0(t))\, n(y,0|t)$$

$$\forall i \geq 1: \quad \frac{\partial n(y,i|t)}{\partial t} - \tilde{k}(t)\frac{\partial(yn(y,i|t))}{\partial y} = -(\alpha_i(t) + \beta_i(t))\, n(y,i|t) + 2\gamma\alpha_{i-1}(t)n(\gamma y, i-1|t), \tag{4.49}$$

with initial conditions

$$i = 0 : n(y,0|0) \equiv n_0(y), \quad \forall i \geq 1 : n(y,i|0) \equiv 0.$$

Furthermore, the common degradation rates models – time-independent decay and Gompertz decay – can be re-parameterized, e.g., $\tilde{k}(t) = c(k_{\max}e^{-k_T t}) = \tilde{k}_{\max}e^{-k_T t}$, resulting in a reduced number of unknowns. This proves that the estimation of the parameter c can be avoided by stating the model in terms of $n(y, i|t)$ as done by (4.49).

Remark 4.12. *As the CFSE concentration, x, and the CFSE induced fluorescence, y, are that closely linked, we will in the following merely study the dynamics and the distribution of y.*

Relation of the CFSE induced fluorescence and the measured fluorescence

While $n(y|t)$ is the distribution of CFSE induced fluorescence and described by the DLSP model (4.49), this model does still not describe the distribution $n(\bar{y}|t)$ of the measured fluorescence, $\bar{Y} = Y + Y_a$. To predict $n(\bar{y}|t)$, the autofluorescence, $p_a(y_a)$, must be taken into account. As $\bar{Y}$ is the sum of CFSE induced fluorescence and autofluorescence, $n(\bar{y}|t)$ is simply the convolution of the label induced fluorescence, $n(y|t)$, and the autofluorescence distribution, $p_a(y_a)$,

$$n(\bar{y}|t) = \int_{\mathbb{R}_+} n(y|t)p_a(\bar{y} - y)dy. \tag{4.50}$$

Hence, the measured fluorescence distribution, $n(\bar{y}|t)$, which is a number density function, can be obtained by simulating (4.35) and computing the convolution integral (4.50).

This is in contrast to the formulation proposed by Thompson (2012), where Y_a is partially contained in the model. If non-trivial noise distributions are studied, the approach by Thompson (2012) requires the repeated simulation of the model. For the example presented in (Banks *et al.*, 2012; Thompson, 2012), one hundred simulations were necessary to obtain a sufficient approximation of the measured distribution $n(\bar{y}|t)$. The rigorous decomposition of the computation of $n(\bar{y}|t)$ in dynamics and measurement overcomes the necessity of repeated model simulations. Independent of the noise distribution, merely a single simulation of the DLSP model is required. Furthermore, this decomposition is in our opinion far more intuitive, as parameters effecting the dynamics and the measurement can be studied separately.

To summarize, in this section the DLSP model has been analyzed in detail. We have shown that for a very general class of division and death rates, $\alpha_i(t)$ and $\beta_i(t)$, the solution of the DLSP model can be computed by solving a system of ODEs. This ODE system has an analytical solution for a rather general class of time independent parameterizations. By determining rigorous error bounds, we furthermore enable the calculation of the required truncation index, S, to achieve a predefined precision.

4.3 Bayesian estimation and uncertainty analysis of proliferation dynamics

In the previous section, we have introduced the DLSP model which is capable of predicting the overall label distribution within the cell population, as well as the properties of subpopulations with common division number. In this section, we propose a Bayesian method to infer the parameters of this model from experimental data.

Remark 4.13. *In the following, we assume that S is chosen large enough to ensure that the approximation error is negligible. As this can be guaranteed using the previously derived error bound, we will not distinguish between the infinite sum, $\sum_{i\in\mathbb{N}_0}$, and the finite sum, $\sum_{i=1}^{S-1}$, to keep the notation simple.*

4.3.1 Bayes' theorem and likelihood function

The unknowns of the proliferation model (4.49) and (4.50) are the time-dependent rates $\alpha_i(t)$, $\beta_i(t)$, and $k(t)$, as well as the initial condition, $n_0(y)$, and the distribution of the autofluorescence, $p_a(y_a)$, together

$$\theta := \left[\{\alpha_i(t)\}_{i=1}^{S}, \{\beta_i(t)\}_{i=1}^{S}, \tilde{k}(t), n_0(y), p_a(y_a)\right]^{\mathrm{T}}. \qquad (4.51)$$

To infer these parameters of the DLSP model from binned snapshot data, $\mathcal{D}$, we use a Bayesian framework, relying on Bayes' theorem,

$$\pi(\theta|\mathcal{D}) = \frac{\mathbb{P}(\mathcal{D}|\theta)\pi(\theta)}{\mathbb{P}(\mathcal{D})}, \qquad (4.52)$$

in which $\pi(\theta)$ is the prior probability of θ, $\mathbb{P}(\mathcal{D}|\theta)$ is the likelihood (or conditional probability) of observing $\mathcal{D}$ given θ, $\pi(\theta|\mathcal{D})$ is the posterior probability of θ given $\mathcal{D}$, and $\mathbb{P}(\mathcal{D}) = \int \mathbb{P}(\mathcal{D}|\theta)\pi(\theta)d\theta$ is the marginal probability of $\mathcal{D}$. As before, we make use of $\mathbb{P}(\mathcal{D})$ being constant given θ, yielding

$$\pi(\theta|\mathcal{D}) \propto \mathbb{P}(\mathcal{D}|\theta)\pi(\theta). \qquad (4.53)$$

In addition, the likelihood can be factorized as the individual binned snapshots are independent,

$$\mathbb{P}(\mathcal{D}|\theta) = \prod_{j=1}^{d_{\mathcal{D}}} \mathbb{P}(\mathcal{SD}_j^{\mathcal{B}}|\theta). \qquad (4.54)$$

To derive the likelihood function $\mathbb{P}(\mathcal{SD}_j^{\mathcal{B}}|\theta)$, note that $\mathcal{SD}_j^{\mathcal{B}}$ can contain two pieces of information: the distribution of the measured label fluorescence encoded in the histogram counts $\bar{H}_{t_j}^l$, and the cell number $\bar{N}_{t_j} = \sum_{l=1}^{d_l} \bar{H}_{t_j}^l$ which might be proportional to the overall population size. While the former information is always available, the latter can only be assessed if the precise handling of the cell culture is reported. As flow cytometry and microscopy are measurement techniques which analyze a certain volume to determine the population size or the cell density, especially the change of medium and the loss of cells due to medium change has to be studied accurately. This is not always feasible, as discussed in (Banks *et al.*, 2011; Thompson, 2012), where large outliers were detected in published datasets.

Given the two pieces of information, $\bar{H}_{t_j}^l$ and $\bar{N}_{t_j}$, the likelihood function becomes

$$\mathbb{P}(\mathcal{SD}_j^{\mathcal{B}}|\theta) = \mathbb{P}(\{\bar{H}_{t_j}^l\}_{l=1}^{d_l}|\bar{N}_{t_j}, \theta)\mathbb{P}(\bar{N}_{t_j}|\theta), \qquad (4.55)$$

in which $\mathbb{P}(\{\bar{H}^l_{t_j}\}_{l=1}^{d_l}|\bar{N}_{t_j},\theta)$ is the likelihood of observing the histogram $\{\bar{H}^l_{t_j}\}_{l=1}^{d_l}$ given that $\bar{N}_{t_j}$ cells are measured, and $\mathbb{P}(\bar{N}_{t_j}|\theta)$ denotes the likelihood that $\bar{N}_{t_j}$ cells are observed.

The likelihood $\mathbb{P}(\bar{N}_{t_j}|\theta)$ strongly depends on the measurement device and the handling of the cells. If, for instance, flow cytometry is used to analyze a fraction f_Ω of the complete volume Ω in which a population of size $N(t,\theta)$ is contained, the number of observed cells is approximately binomially distributed,

$$\mathbb{P}(\bar{N}_{t_j}|\theta) = \frac{N(t,\theta)!}{\bar{N}_{t_j}!(N(t,\theta)-\bar{N}_{t_j})!} f_\Omega^{\bar{N}_{t_j}} (1-f_\Omega)^{N(t,\theta)-\bar{N}_{t_j}}. \tag{4.56}$$

This follows under the assumptions that $N(t,\theta)$ is an integer and that the measurement is a drawing process with replacement. While both assumptions are strictly speaking not correct, for typical experiments with large population size, $N(t,\theta) \gg 1$, and small volume fractions, $f_\Omega \ll 1$, rounding of $N(t,\theta)$ towards the next integer and neglecting the missing replacement results in minor errors. Still, while (4.56) describes the pure statistical properties of the measurement process, other sources of systematic and random error are disregarded, yielding probably an underestimation of the variability (see also discussion in (Thompson, 2012)).

The second contribution to the overall likelihood function, $\mathbb{P}(\{\bar{H}^l_{t_j}\}_{l=1}^{d_l}|\bar{N}_{t_j},\theta)$, is the probability of observing the sampled histogram, $\{\bar{H}^l_{t_j}\}_{l=1}^{d_l}$, of measured fluorescence levels, $\bar{Y}^i_{t_j}$. To evaluate this likelihood for given θ, note that the process of collecting $\{\bar{H}^l_{t_j}\}_{l=1}^{d_l}$ can be modeled as the collection of $\bar{N}_{t_j}$ independent, generalized Bernoulli trials (Grinstead & Snell, 1997). In each trial, d_l outcomes are possible, namely, the observed property of the cell falling into bin 1 ($\bar{Y}^i_{t_j} \in I_1$) up to bin d_l ($\bar{Y}^i_{t_j} \in I_{d_l}$). Given θ the probability of the individual outcomes is

$$p(\bar{Y}^j_{t_j} \in I_l|\theta) = \begin{cases} \int_{I_l} p(\bar{y}|t_j,\theta) d\bar{y} & \text{for } N(t,\theta) > 0 \\ 0 & \text{otherwise,} \end{cases} \tag{4.57}$$

in which $\forall t \in \mathbb{R}_+$ with $N(t,\theta) > 0$:

$$p(\bar{y}|t,\theta) = \frac{n(\bar{y}|t,\theta)}{N(t,\theta)} \tag{4.58}$$

is the probability density of observing the measured output $\bar{Y}_t$ at time t for a given parameter vector θ. The probability mass function of this kind of Bernoulli trials is described by the multinomial distribution (Merran *et al.*, 2000),

$$\mathbb{P}(\{\bar{H}^l_{t_j}\}_{l=1}^{d_l}|\bar{N}_{t_j},\theta) = \frac{\bar{N}_{t_j}!}{\prod_{l=1}^{d_l} \bar{H}^l_{t_j}!} \prod_{i=1}^{m} \left(p(\bar{Y}^i_{t_j} \in I_j|\theta)\right)^{\bar{H}^l_{t_j}}. \tag{4.59}$$

The formulation (4.59) of the likelihood function $\mathbb{P}(\{\bar{H}^l_{t_j}\}_{l=1}^{d_l}|\bar{N}_{t_j},\theta)$ considers merely the statistical properties of the drawing process. However, there are several additional sources of error (Thompson, 2012). It is, for instance, known that data often contain outliers which are not captured by the true data. To obtain a reliable statistical model and to ensure robustness of the parameter estimation, we explicitly consider such outliers by adding a probability distribution of outliers. This results in an altered probability of observing a certain level of fluorescence,

$$\bar{p}(\bar{y}|t,\theta) = \left(1 - \int_{\mathbb{R}_+} p_o(\bar{y}) d\bar{y}\right) \frac{n(\bar{y}|t,\theta)}{N(t,\theta)} + p_o(\bar{y}), \tag{4.60}$$

in which $\int_{\mathbb{R}_+} p_o(\bar{y})d\bar{y}$ is the probability that an observed data point is an outlier and $p_o(\bar{y})$ is the distribution of outliers. The occurrence of outliers might be interpreted as a noise on the population level, requiring a modification of (4.57) and (4.59). If additionally additive measurement noise on the single cell level occurs, this can be included in the autofluorescence distribution, as these two sources are nevertheless indistinguishable.

4.3.2 Efficient evaluation of the likelihood function

Employing the likelihood function introduced above, parameter estimation can be performed. The computational complexity of the parameter estimation task crucially depends on the computational effort associated with the evaluation of the likelihood function. In most studies, the likelihood function for a given parameter θ is evaluated in two steps: At first the model is simulated, and afterwards the distance of model prediction and measurement data is assessed (see, e.g., (Banks *et al.*, 2011)).

To improve the computational efficiency, we intertwine these two steps. Instead of simulating the complete model, we employ the decomposition $n(\bar{y}, i|t, \theta) = p(\bar{y}|i, t, \theta)N(i|t)$ of the distribution of the measured outputs. Merely $N(i|t, \theta)$ is determined by simulation, enabling for the calculation of $\mathbb{P}(\bar{N}_{t_j}|\theta)$, while $p(\bar{y}|i, t, \theta)$ is approximated to enable a direct assessment of $\mathbb{P}(\{\bar{H}^l_{t_j}\}^{d_l}_{l=1}|\bar{N}_{t_j}, \theta)$. This allows us to circumvent the calculation of the convolution integral (4.50), which is the computational bottleneck.

Approximation of measured label distribution

To reduce the computational complexity, we propose an approximation $\hat{n}(\bar{y}|t, \theta)$ of $n(\bar{y}|t, \theta)$ which can be computed without integration. To allow for this approximation, the initial condition is restricted to be a weighted sum of log-normal distributions,

$$n_0(y) = N_0 p_0(y), \quad \text{with } p_0(y) = \sum_{j=1}^{d_f} f_j \log\mathcal{N}(y|\mu^j_0, (\sigma^j_0)^2), \tag{4.61}$$

with fraction parameters $f_j \in [0, 1]$, with $\sum_{j=1}^{d_f} f_j = 1$, and parameters $\mu^j_0, \sigma^j_0 \in \mathbb{R}_+$. The fraction parameters, f_j, determine which fraction of cells belongs to which log-normal distribution. The number of different log-normal distributions is denoted by $d_f \in \mathbb{N}$. In addition, we restrict the measurement noise to be log-normally distributed, $p(y_a) = \log\mathcal{N}(y_a|\mu_a, \sigma^2_a)$. These two assumptions are not restrictive, as any smooth distribution can be approximated arbitrarily well by a sum of log-normal distributions and as autofluorescence levels are known to be approximately log-normally distributed (see, e.g., (Banks *et al.*, 2012; Hawkins *et al.*, 2007; Thompson, 2012)).

Given (4.61), it can be shown that the CFSE induced fluorescence distribution in the subpopulation is

$$n(y, i|t, \theta) = N(i|t)p(y|i, t, \theta), \tag{4.62}$$

with

$$p(y|i, t, \theta) = \sum_{j=1}^{d_f} f_j \underbrace{\log\mathcal{N}(y|\mu^j_i(t, \theta), (\sigma^j_0)^2)}_{=:p_{0,j}(y)} \tag{4.63}$$

and $\mu^j_i(t, \theta) = -i\log(\gamma) - \int_0^t k(\tau)d\tau + \mu^j_0$ (for a proof see Appendix H). This follows directly from the analytical solution of $p(y|i, t, \theta)$. Thus, log-normal distributions are invariant under the considered class of partial differential equations, and log-normal initial conditions result in log-normal label distributions for $t > 0$. This is an important finding, as it holds for a wide range of PDE models, and we denote it as log-normal invariance concept. For the problem at hand, it implies that the fluorescence distribution is a sum of log-normal distributions,

$$n(y|t,\theta) = \sum_{i\in\mathbb{N}_0} n(y,i|t,\theta) = \sum_{i\in\mathbb{N}_0} N(i|t,\theta) \sum_{j=1}^{d_f} f_j \log\mathcal{N}(y|\mu_i^j(t,\theta),(\sigma_0^j)^2). \tag{4.64}$$

By inserting this in the convolution integral (4.50), we obtain by linearity of integration

$$n(\bar{y}|t,\theta) = \int_0^\infty \left(\sum_{i\in\mathbb{N}_0} n(y,i|t,\theta) \log\mathcal{N}(\bar{y}-y|\mu_a,\sigma_a^2)\right) dy \tag{4.65}$$

$$= \sum_{i\in\mathbb{N}_0} N(i|t,\theta) \sum_{j=1}^{d_f} f_j \int_0^\infty \log\mathcal{N}(y|\mu_i^j(t,\theta),(\sigma_0^j)^2) \log\mathcal{N}(\bar{y}-y|\mu_a,\sigma_a^2) dy. \tag{4.66}$$

The individual summands

$$p(\bar{y}|i,t,\theta,p_{0,j}) = \int_0^\infty \log\mathcal{N}(y|\mu_i^j(t,\theta),(\sigma_0^j)^2) \log\mathcal{N}(\bar{y}-y|\mu_a,\sigma_a^2) dy, \tag{4.67}$$

describe the normalized contribution of the j-th log-normal distribution in the initial condition to the measured fluorescence distribution, $n(\bar{y}|t,\theta)$. This can be traced back as the superposition principle holds. Apparently, the efficient assessment of $n(\bar{y}|t,\theta)$ is possible, using an efficient computational scheme for computing $p(\bar{y}|i,t,\theta,p_{0,j})$.

The distribution $p(\bar{y}|i,t,\theta,p_{0,j})$ is the density of the sum of two log-normally distributed random variables, $\bar{Y} = Y + Y_a$, with $Y \sim \log\mathcal{N}(y|\mu_i^j(t,\theta),(\sigma_0^j)^2)$ and $Y_a \sim \log\mathcal{N}(y_a|\mu_a,\sigma_a^2)$. Although similar densities are of interest in many research fields (see, e.g., (Beaulieu, 2004; Fenton, 1960) and references therein), no analytical formula for computing $p(\bar{y}|i,t,\theta,p_{0,j})$ is known. However, several approximations are available. One of the most commonly used approximation has been proposed by Fenton (1960). Fenton employs the fact that although the distribution of the sum of two log-normally distributed random variables is not log-normal, it can still be closely approximated by a log-normal distribution. In (Fenton, 1960), this approximating log-normal distribution is chosen to have the same first two central moments, mean $E_{\bar{y}}^{i,j}(t,\theta)$ and variance $\mathrm{Var}_{\bar{y}}^{i,j}(t,\theta)$, as the actual distribution of the sum.

The time-dependent central moments of $p(\bar{y}|i,t,\theta,p_{0,j})$ are the sums

$$E_{\bar{y}}^{i,j}(t,\theta) = E_y^{i,j}(t,\theta) + E_{y_a}, \tag{4.68}$$

$$\mathrm{Var}_{\bar{y}}^{i,j}(t,\theta) = \mathrm{Var}_y^{i,j}(t,\theta) + \mathrm{Var}_{y_a}, \tag{4.69}$$

of the time-dependent central moments of the label induced fluorescence distribution of the i-th subpopulation of the j-th contribution to the initial condition, $E_y^{i,j}(t,\theta)$ and $\mathrm{Var}_y^{i,j}(t,\theta)$, and the static autofluorescence, E_{y_a} and Var_{y_a}, as it is known from basic statistics (Grinstead & Snell, 1997). These central moments are

$$E_y^{i,j}(t,\theta) = e^{\mu_i^j(t,\theta)} e^{\frac{(\sigma_0^j)^2}{2}}, \tag{4.70}$$

$$\mathrm{Var}_y^{i,j}(t,\theta) = e^{2\mu_i^j(t,\theta)+(\sigma_0^j)^2}\left(e^{(\sigma_0^j)^2} - 1\right), \tag{4.71}$$

for the label distribution and

$$\mathrm{E}_{y_a} = e^{\mu_a} e^{\frac{\sigma_a^2}{2}}, \tag{4.72}$$

$$\mathrm{Var}_{y_a} = e^{2\mu_a+\sigma_a^2}\left(e^{\sigma_a^2} - 1\right), \tag{4.73}$$

for the measurement noise. Following Fenton (1960), the log-normal distribution exhibiting the same overall mean and variance has parameters

$$\hat{\mu}_{\bar{y}}^{i,j}(t,\theta) = \log(\mathrm{E}_{\bar{y}}^{i,j}(t,\theta)) - \frac{1}{2}\log\left(\frac{\mathrm{Var}_{\bar{y}}^{i,j}(t,\theta)}{\mathrm{E}_{\bar{y}}^{i,j}(t,\theta)} + 1\right), \tag{4.74}$$

$$\hat{\sigma}_{\bar{y}}^{i,j}(t,\theta) = \sqrt{\log\left(\frac{\mathrm{Var}_{\bar{y}}^{i,j}(t,\theta)}{\mathrm{E}_{\bar{y}}^{i,j}(t,\theta)} + 1\right)}, \tag{4.75}$$

yielding the approximation

$$\hat{p}(\bar{y}|i,t,\theta,p_{0,j}) = \log\mathcal{N}(\bar{y}|\hat{\mu}_{\bar{y}}^{i,j}(t,\theta),(\hat{\sigma}_{\bar{y}}^{i,j}(t,\theta))^2) \tag{4.76}$$

of $p(y|i,t,\theta,p_{0,j})$. Own studies revealed (not shown), that this approximation is in many cases indistinguishable from the true distribution. In particular, if one of the distributions becomes narrow, the approximation can be made arbitrarily good. This is helpful, as the precise parameterization of the initial condition might be a degree of freedom, which can be used to regulate the approximation quality. A detailed assessment of the approximation quality may be found in (Fenton, 1960).

Given the approximation $\hat{p}(\bar{y}|i,t,\theta,p_{0,j})$, the approximation

$$\hat{n}(\bar{y}|t,\theta) = \sum_{i\in\mathbb{N}_0} N(i|t,\theta) \sum_{j=1}^{d_f} f_j \log\mathcal{N}(\bar{y}|\hat{\mu}_{\bar{y}}^{i,j}(t,\theta),(\hat{\sigma}_{\bar{y}}^{i,j}(t,\theta))^2) \tag{4.77}$$

of the measured fluorescence distribution can be computed. This approximation is the sum of log-normal distributions for which the parameters can be calculated analytically. Therefore, it merely requires the evaluation of the log-normal distribution at different points, which can be made fairly efficient using lookup tables. The approximation (4.77) can be computed orders of magnitudes faster than the actual convolution integral (4.50).

Calculation of the likelihood function using only partial simulations

The approximation introduced above allows for the efficient assessment of the fluorescence distribution, $n(\bar{y}|t,\theta)$, as well as the probability density of observing different label concentrations in the presence of outliers,

$$\hat{\bar{p}}(\bar{y}|t,\theta) = \left(1 - \int_{\mathbb{R}_+} p_o(\bar{y})d\bar{y}\right) \sum_{i\in\mathbb{N}_0} \frac{N(i|t,\theta)}{N(t,\theta)} \sum_{j=1}^{d_f} f_j \log\mathcal{N}(\bar{y}|\hat{\mu}_{\bar{y}}^{i,j}(t,\theta),(\hat{\sigma}_{\bar{y}}^{i,j}(t,\theta))^2) + p_o(\bar{y}). \tag{4.78}$$

However, to evaluate the likelihood functions $\mathbb{P}(\{\bar{H}_{t_j}^l\}_{l=1}^{d_l}|\bar{N}_{t_j},\theta)$ according to (4.59), the probability that a measurement $\bar{Y}_{t_j}^j$ is contained in bin j, $p(\bar{Y}_{t_j}^j \in I_l|\theta)$, is required. Using the ad hoc formulation (4.57), the evaluation of $\mathbb{P}(\{\bar{H}_{t_j}^l\}_{l=1}^{d_l}|\bar{N}_{t_j},\theta)$ requires a large number of integrations over $\hat{p}(\bar{y}|t,\theta)$. To circumvent this limiting step, the structure of $\hat{n}(\bar{y}|t,\theta)$ is employed. Note that the probability that $\bar{Y} < \bar{y}_l$ is observed for a cell which originated from the j-th summand of the initial condition, $p_{0,j}(y)$, and which has undergone i cell divisions is approximately

$$\hat{\Phi}^{i,j}(\bar{y}_l|t,\theta) = \int_0^{\bar{y}_l} \log \mathcal{N}(\bar{y}|\hat{\mu}_{\bar{y}}^{i,j}(t,\theta),(\hat{\sigma}_{\bar{y}}^{i,j}(t,\theta))^2)d\bar{y} \tag{4.79}$$

$$= \frac{1}{2}\text{erfc}\left(-\frac{\log(\bar{y}_l) - \hat{\mu}_{\bar{y}}^{i,j}(t,\theta)}{\sqrt{2}\hat{\sigma}_{\bar{y}}^{i,j}(t,\theta)}\right). \tag{4.80}$$

This follows from the approximation (4.76) using that the cumulative probability of a log-normal distribution is the error function, erfc(·). It shows that integration can be carried out analytically.

To exploit this, (4.57) is reformulated in terms of (4.80), yielding

$$\hat{p}(\bar{Y}_{t_j}^j \in I_l|\theta) = (1 - P_o(\bar{y} \in \mathbb{R}_+)) \sum_{i\in\mathbb{N}_0} \frac{N(i|t,\theta)}{N(t,\theta)} \sum_{j=1}^{d_f} f_j(\hat{\Phi}^{i,j}(\bar{y}_{l,\text{ub}}|t,\theta) - \hat{\Phi}^{i,j}(\bar{y}_{l,\text{lb}}|t,\theta)) + P_o(\bar{y} \in I_l), \tag{4.81}$$

an expression which is free of parameter dependent integrals. Hence, the assessment of $\hat{p}(\bar{Y}_{t_j}^j \in I_l|\theta)$ only requires the simulation of an ODE to determine $N(i|t)$, the calculation of $\hat{\mu}_{\bar{y}}^{i,j}(t,\theta)$ and $\hat{\sigma}_{\bar{y}}^{i,j}(t,\theta)$ – for both analytical solution exist for common choices of $k(t)$ –, and the repeated evaluation of the error function. The latter can be made computationally efficient using lookup tables.

Furthermore, usage of (4.81) for the calculation of the likelihood function $\mathbb{P}(\{\bar{H}_{t_j}^l\}_{l=1}^{d_l}|\bar{N}_{t_j},\theta)$ renders the explicit computation of the distribution of label induced fluorescence or measured fluorescence pointless. By combining (4.81) with the state truncation, $i < S$, a computational scheme for the approximation of the likelihood function can be derived. The corresponding pseudocode in depicted in Algorithm 4.1, which is efficient with respect to computation time and storage requirements.

Remark 4.14. *In the remainder of this section, we employ the approximation $\hat{p}(\bar{Y}_{t_j}^j \in I_l|\theta)$ of $p(\bar{Y}_{t_j}^j \in I_l|\theta)$. This approximation uses the truncation of the population model at $i < S$ and the approximation of the distribution of the sum $\bar{Y} = Y + Y_a$. As the truncation error associated with S can be controlled using the error bounds, and the approximation error of the sum of two logarithmically distributed variables can be reduced by re-parametrization of the initial condition (see discussion above), this approximation can be made arbitrarily precise.*

4.3.3 Parameterization of unknown functions

After introducing an efficient approach to assess the likelihood of the model to generate the data, the problem of parameter estimation can be approached. Therefore, the unknown functions have to be parameterized. In the following, we will discuss different parametrization approaches, as well as methods to reduce the number of parameters.

Autofluorescence Several studies (see, e.g., (Thompson, 2012)) showed that in many cell systems, such as T lymphocytes, the levels of autofluorescence are log-normally distributed. Therefore, we choose $p_a(y_a) = \log \mathrm{N}(y_a|\mu_a,\sigma_a^2)$ (4.2). As μ_a and σ_a^2 can be computed from

Algorithm 4.1 Pseudocode for likelihood function evaluation.

Require: Parameter vector $\theta = \left[\{\alpha_i(t)\}_{i=1}^{S}, \{\beta_i(t)\}_{i=1}^{S}, k(t), N_0(x), p_a(y_a)\right]^{\mathrm{T}}$, bounds of the interval $\bar{y}_{l,\mathrm{lb}}$ and $\bar{y}_{l,\mathrm{ub}}$, truncation index S.
Solve model (4.25) for subpopulation size, $N(i|t, \theta)$, with $i \in \{0, \ldots, S-1\}$.
Compute size of overall population, $N(t, \theta) = \sum_{i=0}^{S-1} N(i|t, \theta)$.
Compute probability of finding any outliers, $P_o(\bar{Y}_{t_j}^{j} \in \mathbb{R}_+) = \int_{\mathbb{R}_+} p_o(\bar{y}) d\bar{y}$.
for $l = 1$ to d_I **do**
  Compute probability of finding outliers in I_l, $P_o(\bar{Y}_{t_j}^{j} \in I_l) = \int_{I_l} p_o(\bar{y}) d\bar{y}$.
end for
for $j = 1$ to $d_{\mathcal{D}}$ **do**
  Compute likelihood of observing the $\bar{N}_{t_j}$ cells at t_j using (4.56), $\mathbb{P}(\bar{N}_{t_j}|\theta)$.
  for $i = 1$ to $S - 1$ **do**
    Evaluate equation (4.74), $\hat{\mu}_{\bar{y}}^{i,j}(t_j, \theta)$.
    Evaluate equation (4.75), $\hat{\sigma}_{\bar{y}}^{i,j}(t_j, \theta)$.
    for $l = 1$ to d_I **do**
      Evaluate cumulative log-normal distributions at $\bar{y}_{l,\mathrm{lb}}$, $\hat{\Phi}^{i,j}(\bar{y}_{l,\mathrm{lb}}|t_j, \theta)$.
      Evaluate cumulative log-normal distributions at $\bar{y}_{l,\mathrm{ub}}$, $\hat{\Phi}^{i,j}(\bar{y}_{l,\mathrm{ub}}|t_j, \theta)$.
    end for
    Compute probability of observing bin $\bar{Y}_{t_j}^{j} \in I_l$ using (4.81), $\hat{p}(\bar{Y}_{t_j}^{j} \in I_l|\theta)$.
  end for
  Compute likelihood of observing the j-th histogram using (4.59), $\mathbb{P}(\{\bar{H}_{t_j}^{l}\}_{l=1}^{d_I}|\bar{N}_{t_j}, \theta)$.
  Compute likelihood of observing the j-th snapshot, $\mathbb{P}(\mathcal{SD}_j^{\mathcal{B}}|\theta) = \mathbb{P}(\{\bar{H}_{t_j}^{l}\}_{l=1}^{d_I}|\bar{N}_{t_j}, \theta)\mathbb{P}(\bar{N}_{t_j}|\theta)$.
end for
Compute overall likelihood, $\mathbb{P}(\mathcal{D}|\theta) = \prod_{j=1}^{d_{\mathcal{D}}} \mathbb{P}(\mathcal{SD}_j^{\mathcal{B}}|\theta)$.

experiments with unlabeled cells, they might be estimated independently of the other parameters. However, if appropriate control experiments are not available, μ_a and σ_a^2 must be included in the estimation problem.

Initial fluorescence distribution To ensure computational efficiency, $n_0(y)$ is modeled as a weighted sum of log-normal distributions (4.61). The parameters of the initial values $n_0(y)$, N_0 and $\{f_j, \mu_0^j, (\sigma_0^j)^2\}_{j=1}^{d_f}$, can either be inferred during the main loop of the parameter estimation, or beforehand, from the available data for $t = 0$. While the former approach is statistically more rigorous – all measurement data are handled similarly –, the uncertainty in the initial condition is in general negligible, which justifies the latter procedure. Furthermore, the separate estimation of the initial condition is in general more efficient, as a large estimation problem is split up into two smaller ones, resulting in a reduction of the overall complexity.

Label degradation rate As discussed in Section 4.2.1, $\tilde{k}(t)(= ck(t))$ merely depends on $P(S_i)$, which encodes the binding properties of CFSE to proteins with different half-life times. Hence, a parametrization of $\tilde{k}(t)$ can be circumvented by using a parametric model for $P(S_i)$. The analysis of half-life times carried out by Eden *et al.* (2011) suggests that a Poisson distribution might be a reasonable choice. However, for each likelihood function evaluation, $k(t)$

has to be computed from $P(S_i)$, which might be computationally demanding. This renders a direct parameterization of $\tilde{k}(t)$, using, e.g., the Gompertz decay model, $\tilde{k}(t) = \tilde{k}_{\max} e^{-k_T t}$, a better choice if $P(S_i)$ is not of direct interest. The Gompertz decay model captures the key properties of the decay process (see discussion in Section 4.2.1 and (Banks *et al.*, 2011)) and has only two parameters, which avoid over-parameterization.

Division and death rates Proliferation assays are mainly used to study the dynamics of cell division and cell death. This renders the appropriate parameterization of $\alpha_i(t)$ and $\beta_i(t)$ essential. While in most studies tailored ansatz functions are employed to fit the experiment data using a minimal number of parameters (see, e.g., (Banks *et al.*, 2011, 2010, 2012)), we argue that highly flexible parameterizations are required. If the degrees of freedom are too limited, the estimation result is heavily influenced by the parameterization. Therefore, the achieved fit might be far from being optimal, prohibiting a statistically meaningful model rejection. Furthermore, it is difficult to gain new insight, and – even worse – if the parameterization of $\alpha_i(t)$ and $\beta_i(t)$ is not flexible enough, the parameter uncertainties will be underestimated.

Remark 4.15. *In a Bayesian framework under-parameterization of functions is at least as disadvantageous as over-parameterization. While over-parameterization manifests itself in easily detectable codependencies between parameters, under-parameterization is indistinguishable from an incorrect model structure.*

In the literature, mainly splines are used to model cell division and cell death rates (Banks *et al.*, 2011, 2010, 2012; Luzyanina *et al.*, 2007a, 2009; Thompson, 2012). Splines $S(t)$ are piecewise polynomial functions (De Boor, 1978),

$$S(t) = \begin{cases} \mathrm{Poly}_1(t) & \text{for } \tau_0 \le t < \tau_1, \\ \mathrm{Poly}_2(t) & \text{for } \tau_1 \le t < \tau_2, \\ \vdots & \\ \mathrm{Poly}_{d_P}(t) & \text{for } \tau_{d_{\mathrm{Poly}}-1} \le t \le \tau_{d_{\mathrm{Poly}}}, \end{cases} \tag{4.82}$$

generally subject to continuity constraints, $\mathrm{Poly}_j(\tau_j) = \mathrm{Poly}_{j+1}(\tau_j)$ for $i = 1, \ldots, d_{\mathrm{Poly}} - 1$. The advantage of splines compared to, e.g., higher-order polynomials, is that oscillations at the boundary of the domains, also known as Runge's phenomenon (Runge, 1901), are reduced.

In practice, mainly first and third order splines are used. Given a set of nodes, $\{\tau_j\}_{i=0}^{d_{\mathrm{Poly}}}$, the only free parameters of first order splines are the heights $\{s_j = S(\tau_j)\}_{i=0}^{d_{\mathrm{Poly}}}$ (see Figure 4.6). This makes it easy to ensure positivity of $S(t)$, which is required for $\alpha_i(t)$ and $\beta_i(t)$. The spline is positive, $\forall t \in [\tau_0, \tau_{d_{\mathrm{Poly}}}] : S(t) \geq 0$, if and only if $\forall i : s_i \geq 0$. Handling of positivity constraints is more difficult for higher dimensional splines, as positivity at the nodes is not longer sufficient for positivity of $S(t)$ in $t \in [\tau_0, \tau_{d_{\mathrm{Poly}}}]$. However, by parametrizing $\log(\alpha_i(t))$ and $\log(\beta_i(t))$ using splines, instead of $\alpha_i(t)$ and $\beta_i(t)$, we can also overcome this problem.

Due to the high-degree of flexibility, splines are in our opinion the best choice to model division and death rates. Still, it has to be made sure that the number of nodes is large enough to avoid under-parameterization. The spline coefficients of $\alpha_i(t)$ and $\beta_i(t)$ are in the following denoted by $\{a_i^j\}_{j=1}^{d_{a_i}}$ and $\{b_i^j\}_{j=1}^{d_{b_i}}$, respectively.

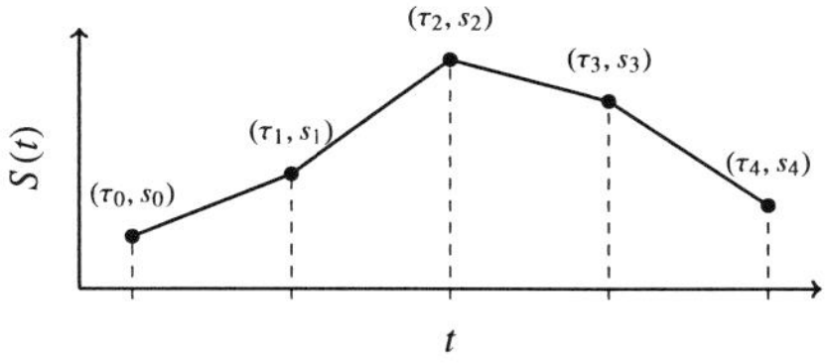

Figure 4.6: Illustration of first order spline.

4.3.4 Bayesian parameter estimation and uncertainty analysis

Given Algorithm 4.1 and a parameterization of the unknowns, the posterior distribution $\pi(\theta|\mathcal{D})$ of the parameters

$$\theta := \Big[\underbrace{\big\{\{a_i^j\}_{j=1}^{d_{a_i}}\big\}_{i=1}^{S}}_{\rightarrow\alpha_i(t)}, \underbrace{\big\{\{b_i^j\}_{j=1}^{d_{b_i}}\big\}_{i=1}^{S}}_{\beta_i(t)}, \underbrace{\tilde{k}_{\max}, k_T}_{k(t)}, \underbrace{N_0, \{f_j, \mu_0^j, (\sigma_0^j)^2\}_{j=1}^{d_f}}_{n_0(y)}, \underbrace{\mu_a, \sigma_a^2}_{p_a(y_a)}\Big]^{\mathrm{T}} \tag{4.83}$$

can be analyzed. This posterior distribution encodes the available knowledge about the parameters. Its maximum defines the maximum a posteriori estimate, θ^*, and the shape contains information about the parameter and prediction uncertainties (see discussion in Section 2).

Annealed MCMC sampling of posterior distribution

Similar to most inference problems, the structure of the posterior distribution $\pi(\theta|\mathcal{D})$ is unknown and prohibits direct sampling. While in Chapter 3 we were able to establish unimodality of the respective posterior distribution arising for the estimation of parameter heterogeneity in signaling pathways, this is not possible here. The distribution $\pi(\theta|\mathcal{D})$ might possess multiple modes, which renders efficient MCMC sampling, defined in terms of statistically independent sample members per unit of time, challenging.

To approach this problem, we employ a combination of annealing MCMC methods (Doucet *et al.*, 2000; Geyer & Thompson, 1995; Neal, 2001; Vyshemirsky & Girolami, 2008) and adaptive Metropolis with delayed rejection (Haario *et al.*, 2006, 2001). Instead of sampling directly the posterior distribution, a sequence of distributions is defined,

$$\pi_{T_j}(\theta|\mathcal{D}) \propto (\mathbb{P}(\mathcal{D}|\theta))^{T_j}\pi(\theta), \quad 0 \leq T_1 < \cdots < T_j < \cdots T_{d_T} = 1, \tag{4.84}$$

with temperature parameters T_j. This sequence bridges the prior distribution $\pi_0(\theta|\mathcal{D})$ and the posterior distribution $\pi_1(\theta|\mathcal{D})$. By drawing a sample from $\pi_{T_j}(\theta|\mathcal{D})$ and using the result to initialize the sampling of $\pi_{T_{j+1}}(\theta|\mathcal{D})$, strong dependence on the initial condition and trapping in local maxima can be avoided, which yields higher effective sampling rates (Doucet *et al.*, 2000; Geyer & Thompson, 1995). This annealing approach is complemented by the application of adaptive methods to sample the individual distribution (Haario *et al.*, 2006), and the reuse of the adapted kernel in the next sampling step.

While annealing MCMC methods with sophisticated update steps enable the computation of a representative sample, $\mathcal{S} = \{\theta^k\}_{k=1}^{d_S}$ from $\pi(\theta|\mathcal{D})$, the sampling speed can be increased further by re-parameterization of θ. In particular, sampling of φ, with $\theta = e^{\varphi}$, increases the sampling speed significantly due to improved numerical conditions. Additionally, sampling of φ ensures positivity of the entries of θ.

Parameter and prediction uncertainties

Given a representative sample, $\mathcal{S} = \{\theta^k\}_{k=1}^{d_S}$, parameter and prediction uncertainties can be determined. Similar to Section 2.3.2, the $100(1-\alpha)\%$ Bayesian confidence intervals for the parameters is

$$\forall j: \quad [\theta_j]^{(1-\alpha)} = [\theta_j^{(\alpha/2)}, \theta_j^{(1-\alpha/2)}], \tag{4.85}$$

with $\theta_j^{(\alpha)}$ being the 100α-th percentile of $\pi(\theta_i|\mathcal{D})$. The prediction uncertainty on the other hand is, dependent on the property of interest, a function of time and/or label properties (see Section 3.3.5). In the following, we study the confidence interval of $n(\bar{y}|t,\theta)$,

$$\forall t, \bar{y} \in \mathbb{R}_+ : \quad [n]^{(1-\alpha)}(\bar{y}|t) = [n^{(\alpha/2)}(y|t), n^{(1-\alpha/2)}(\bar{y}|t)], \tag{4.86}$$

and $N_i(\bar{y}|t,\theta)$,

$$\forall i \in \mathbb{N}_0, t \in \mathbb{R}_+ : \quad [N_i]^{(1-\alpha)}(t) = [N_i^{(\alpha/2)}(t), N_i^{(1-\alpha/2)}(t)]. \tag{4.87}$$

The confidence intervals for other model properties can be defined accordingly.

Computation of maximum a posteriori estimate

Besides the evaluation of the global properties of $\pi(\theta|\mathcal{D})$, also the maximum a posteriori estimate is important, among others for model comparison or model rejection. To compute θ^* the nonlinear optimization problem

$$\underset{\theta \in \mathbb{R}_+^{d_\varphi}}{\text{maximize}} \quad \mathbb{P}(\mathcal{D}|\theta)\pi(\theta) \tag{4.88}$$

has to be solved. This can be done using global optimization methods, however, given the sample $\mathcal{S} = \{\theta^k\}_{k=1}^{d_S}$, local methods are sufficient. The parameter $\theta^{*,\mathcal{S}}$, with $\mathbb{P}(\mathcal{D}|\theta^{*,\mathcal{S}})\pi(\theta^{*,\mathcal{S}}) \geq \mathbb{P}(\mathcal{D}|\theta^k)\pi(\theta^k)$ for $k \in \{1,\ldots,d_S\}$, is in general close to the global optimum of the posterior distribution. Hence, local optimization starting at $\theta^{*,\mathcal{S}}$ will in general provide a reliable approximation of the maximum a posteriori estimate. In our experience, this procedure is at least as robust as direct global optimization.

4.4 Application example: Proliferation of T lymphocytes

In response to bacterial and viral infection, the human body produces a multitude of immune cells, among others, T lymphocytes (Janeway *et al.*, 2001). T lymphocytes belong to the adaptive immune system and their pathogen recognition improves over time to allow for a highly specific response. To successfully fight infection, it is important that the pool of T lymphocytes and other immune cell expands fast enough. This renders the study of T lymphocytes proliferation and its dependence on different stimuli interesting (Bird *et al.*, 1998; Hawkins *et al.*, 2007).

In this section, we consider proliferation assay data collected in (Luzyanina *et al.*, 2007b), after stimulation of CFSE labeled cells with antibodies against CD3 and CD28 receptors. These antibodies induce cell proliferation, which is in the following analyzed using the division- and label-structured population models. To allow for the comparison of our results with existing methods, we employed a dataset which has previously been analyzed using division-structured population models (Luzyanina *et al.*, 2007a) and label-structured population models (Banks *et al.*, 2011, 2010; Luzyanina *et al.*, 2009, 2007b).

The goal of our study is to analyze the performance of the proposed approach. Furthermore, different model hypotheses will be compared and the parameter and prediction uncertainty of the best model will be analyzed in detail.

4.4.1 Measurement data and error model

The available measurement data are CFSE histograms collected on the day of stimulation (day 0) and the five subsequent days (day 1 to day 5). In the remainder of this section, merely the data observed on day 1, day 2, day 4, and day 5 are used for parameter estimation (similar to the study in (Banks *et al.*, 2012)). The other observations (day 0 and day 3) seem to be corrupted by measurement noise, as discussed by Luzyanina *et al.* (2007b) and Banks *et al.* (2011).

Unfortunately, the removal of the first measurement (day 0) results in non-identifiability of $\alpha_i(t)$ and $\beta_i(t)$ for $t \in [0,1]$ [d], and prevents the direct estimation of the initial condition at $t = 0$ [d], $n_0(y)$, from the measurement data. To circumvent both problems, we consider only the time in between the earliest and the latest used measurement in the reduced dataset, $t \in [1,5]$ [d]. Accordingly, the initial condition is the distribution of label induced fluorescence at $t = 1$ [d]. This initial condition is parametrized using a weighted sum of three log-normal distributions, and the corresponding distribution parameters are estimated from the histogram measured on day 1. An analysis of the uncertainty of $n_0(y)$ reveals that it is negligible, which is why it is not taken into account.

To compare the remaining data with the model predictions, we use

$$\mathbb{P}(\bar{N}_{t_j}|\theta) = \mathcal{N}(N(t,\theta)|\bar{N}_{t_j}, (0.01 \cdot \bar{N}_{t_j})^2) \tag{4.89}$$

and $\mathbb{P}(\{\bar{H}^l_{t_j}\}_{l=1}^{d_l}|\bar{N}_{t_j},\theta)$ as in (4.59), with a total outlier probability of 5%, $\int_{\mathbb{R}_+} p_o(\bar{y})d\bar{y} = 0.05$, which is split equally among all bins. The likelihood function $\mathbb{P}(\bar{N}_{t_j}|\theta)$ has been adopted, as the pure sampling error described by (4.56) seems to underestimate the variability in the measurement of the overall population. This is also supported by the detection of the two outliers (day 0 and day 3). In the following, we assume that the measurement error for the population error is normally distributed and proportional to the populations size, with a relative standard deviation of one percent. While the error model might be questioned, no better one is available in literature. To obtain a more precise error model, additional experiments and the analysis of noise sources are necessary.

4.4.2 Model hypotheses

The precise proliferation dynamics of T lymphocytes are poorly understood. In the literature, different hypotheses have been formulated, including: constant cell division and death rates (Revy *et al.*, 2001); time- and division number-dependent cell division and death rates (De Boer *et al.*, 2006; Luzyanina *et al.*, 2007a); and time- and label concentration dependent cell division and death rates (Banks *et al.*, 2011, 2010; Luzyanina *et al.*, 2009). While the latter are known to provide a good fit for the considered dataset, it is widely accepted that optimized labeling does not alter proliferation rates (Hawkins *et al.*, 2007). Hence, the underlying assumption is biologically inconsistent.

We presume that the dependence of the proliferation rate on the label concentration is only required as the LSP model employed in the respective studies cannot account for popula-

Table 4.1: Lower and upper bounds for model parameters. The unit of time is hour.

parameter	a_i^j	b_i^j	$\tilde{k}_{\max}$	k_T	μ_a	σ_a^2
lower bound	10^{-3}	10^{-3}	10^{-2}	10^{-4}	10^{-2}	10^{-2}
upper bound	10	10	10	10	10	10

tion heterogeneity induced by different division numbers. To check this hypothesis, and to perform an in-depth analysis of α and β, we compare four model alternatives:

- $\mathcal{M}_1$: α and β are constant.
- $\mathcal{M}_2$: α and β are only time-dependent.
- $\mathcal{M}_3$: α and β are only division number-dependent.
- $\mathcal{M}_4$: α and β are time- and division number-dependent.

As the last model, $\mathcal{M}_4$, is highly complex, we also consider two simplified versions:

- $\mathcal{M}_{4,\alpha(t)}$: α is time- and division number-dependent, β is only division number-dependent.
- $\mathcal{M}_{4,\beta(t)}$: α is only division number-dependent, β is time- and division number-dependent.

To model the time dependence of α and β, linear splines with six equidistant nodes are used. A fine spacing of nodes did not result in significant improvement.

To ensure an unbiased analysis of the proliferation dynamics, we avoid the utilization of prior information. We merely restrict the parameters to a biologically plausible regime, consistent with earlier studies. The lower and upper bounds for the individual parameters are provided in Table 4.1.

4.4.3 Parameter estimation and model comparison

To estimate the parameters of the models, at first, the posterior distributions are explored using MCMC sampling. The best parameter in each sample is, in a second step, used as starting point of a local minimization via the `MATLAB` routine `fmincon`. This yields a maximum a posteriori estimate θ^* for each model.

For the comparison of model predictions and measurement data, the measured histograms $\{\bar{H}_{t_j}^l\}_{l=1}^{d_l}$ and the predicted histograms

$$\left\{H_{t_j}^l = \int_{I_l} n(\bar{y})|t_j, \theta)d\bar{y}\right\}_{l=1}^{d_l}, \tag{4.90}$$

are depicted in Figure 4.7 and 4.8. It is apparent that $\mathcal{M}_1$, $\mathcal{M}_2$, and $\mathcal{M}_3$ fail to fit the data. In particular, the distributions observed on day 2 and day 4 cannot be reproduced accurately. On the other hand, $\mathcal{M}_4$, which possesses time- and division number-dependent cell division and cell death rates, achieves a good agreement with the measurement data. This proves our presumption that label dependent proliferation rates are unnecessary, if the population heterogeneity induced by different division number is taken into account.

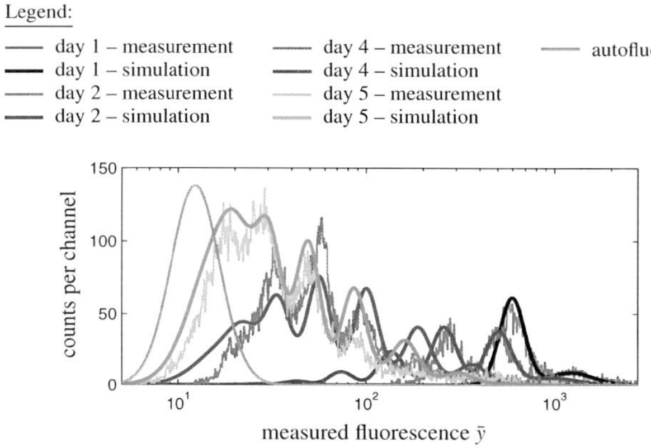

(a) Best fit of model $\mathcal{M}_1$, $\alpha_i(t) = \alpha$ and $\beta_i(t) = \beta$.

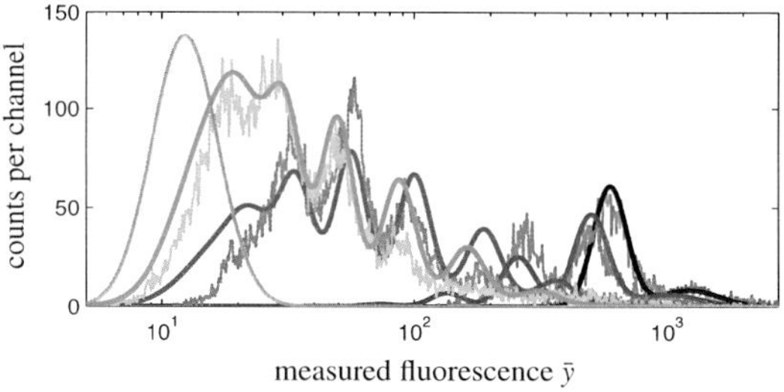

(b) Best fit of model $\mathcal{M}_2$, $\alpha_i(t) = \alpha(t)$ and $\beta_i(t) = \beta(t)$.

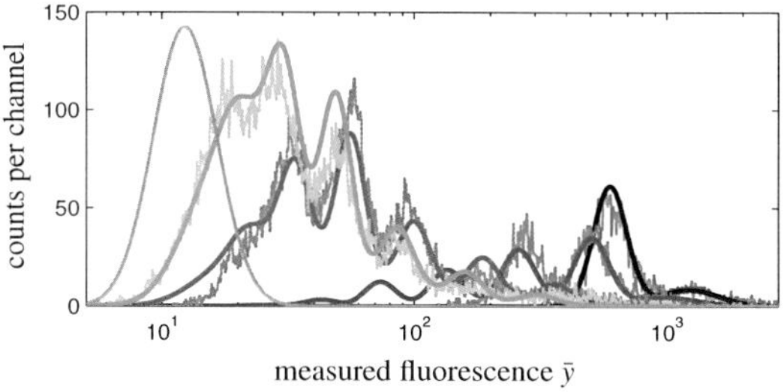

(c) Best fit of model $\mathcal{M}_3$, $\alpha_i(t) = \alpha_i$ and $\beta_i(t) = \beta_i$.

Figure 4.7: Comparison of measurement data and fitted solution of: (a) model $\mathcal{M}_1$, (b) model $\mathcal{M}_2$, and (c) model $\mathcal{M}_3$. For simulation, the maximum a posterior estimate has been used.

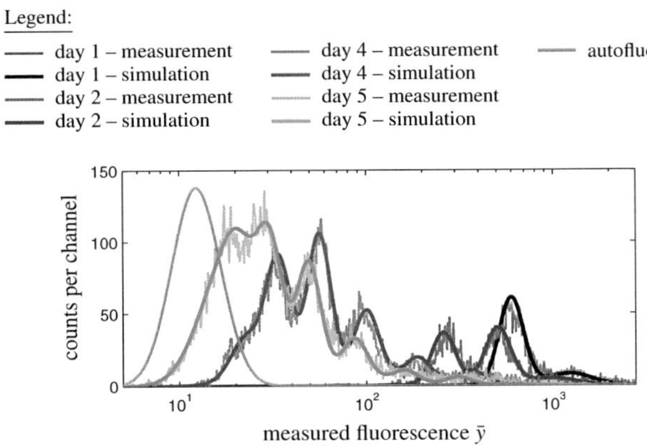

(a) Best fit of model $\mathcal{M}_{4,\alpha(t)}$, $\alpha_i(t)$ and $\beta_i(t) = \beta_i$.

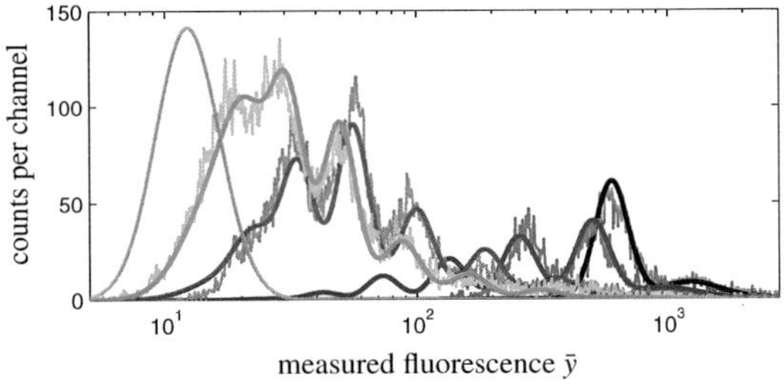

(b) Best fit of model $\mathcal{M}_{4,\beta(t)}$, $\alpha_i(t) = \alpha_i$ and $\beta_i(t)$.

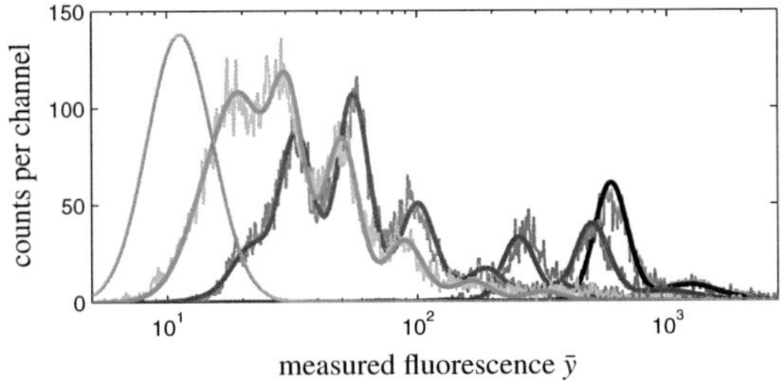

(c) Best fit of model $\mathcal{M}_4$, $\alpha_i(t)$ and $\beta_i(t)$.

Figure 4.8: Comparison of measurement data and fitted solution of: (a) model $\mathcal{M}_{4,\alpha(t)}$, (b) model $\mathcal{M}_{5,\beta(t)}$, and (c) model $\mathcal{M}_4$. For simulation, the maximum a posterior estimate has been used.

The study of the fits achieved by $\mathcal{M}_{4,\alpha(t)}$ and $\mathcal{M}_{4,\beta(t)}$ reveals that the time and division number dependence of the cell division rates seems to be crucial – $\mathcal{M}_{4,\alpha(t)}$ fits the data –, while division number dependence of the death rates is not sufficient. This is also supported by the model comparison using the Bayesian information criterion (BIC) (Schwarz, 1978),

$$\mathrm{BIC} = -2\log\left(\pi(\theta|\mathcal{D})\right) + d_\theta \log\left(\sum_{j=1}^{d_\mathcal{D}} \bar{N}_{t_j}\right). \tag{4.91}$$

The BIC value calculated for $\mathcal{M}_{4,\alpha(t)}$ is lowest (see Table 4.2), though the achieved likelihood value of model $\mathcal{M}_4$ is lower. This indicates that model $\mathcal{M}_4$ is overly complex.

4.4.4 Parameter and prediction uncertainty

To gain further insight into the proliferation process, $\mathcal{M}_{4,\alpha(t)}$ has been analyzed in detail. Therefore, the MCMC sampling is run till the chain converged (geweke value of 0.7), which took several weeks on a standard machine. It turns out, that the parameters $\tilde{k}_{\max}$, k_T, μ_a, σ_a^2, and $\{\beta_i\}_{i=1}^S$ can be estimated quite precisely, while many entries of $\{\{a_i^j\}_{j=1}^{d_{a_i}}\}_{i=1}^S$ are highly uncertain (results not shown).

Interestingly, the uncertainty in $\{\{a_i^j\}_{j=1}^{d_{a_i}}\}_{i=1}^S$ is highly structured. For low division numbers ($i = 0$ and $i = 1$), coefficients a_i^j corresponding to nodes of the spline with $\tau_j > 3$ [d] are uncertain. In contrary, for higher division numbers ($i = 4$ and $i = 5$), coefficients a_i^j corresponding to nodes of the spline with $\tau_j < 2$ [d] are uncertain. This finding can be explained easily by noticing that $N(0|t,\theta)$ and $N(1|t,\theta)$ is small for $t > 3$, while $N(4|t,\theta)$ and $N(5|t,\theta)$ is small for $t < 2$ (Figure 4.9). Hence, the uncertain coefficients are multiplied by small numbers – the effective division rate is $\alpha_i(t)N(i|t,\theta)$ – and therefore do not have any effect on the model response. Beyond the identifiability problems for $\{\{a_i^j\}_{j=1}^{d_{a_i}}\}_{i=1}^S$, this also explains the small prediction uncertainties isn the subpopulation sizes, $N(i|t,\theta)$ (Figure 4.9), and the effective division rate, $\alpha_i(t)N(i|t,\theta)$ (Figure 4.10).

Besides the tight confidence intervals for $N(i|t,\theta)$ and $\alpha_i(t)N(i|t,\theta)$, Figure 4.9 and Figure 4.10 show that the estimated effective division rates, $\alpha_i(t)N(i|t,\theta)$, are always slightly delayed compared to the size of the subpopulation $N(i|t,\theta)$. This seems to result in the best agreement of model predictions and data. Biologically, this delay might be interpreted as a minimal time between two divisions. Though this is biologically plausible, it has not been observed in previous studies. The reason might be that such unexpected findings crucially rely on a flexible parameterization of the model, which in turns requires Bayesian methods to study uncertainties and to determine valid predictions.

Due to the complexity of T lymphocytes proliferation and proliferation assays, such sophisticated Bayesian methods have not been available so far. The main limitation has been the computation effort associated to the simulation and the evaluation of the likelihood function. Our approach reduced the computational complexity, which enabled the exploration of the posterior distribution. However, despite the speed-up, the computational effort remains large as the parameter space is high-dimensional. This shows that the proposed modeling and likelihood function evaluation scheme is promising, but improved MCMC samples are required to ensure faster mixing of the chain.

Table 4.2: Comparison of model alternatives and the achieved value of the Bayesian information criterion. The best value is obtained for model $\mathcal{M}_{4,\alpha(t)}$, which is highlighted.

properties	$\mathcal{M}_1$	$\mathcal{M}_2$	$\mathcal{M}_3$	$\mathcal{M}_{4,\alpha(t)}$	$\mathcal{M}_{4,\beta(t)}$	$\mathcal{M}_4$
division number-dependent α	–	–	✓	✓	✓	✓
time-dependent α	–	✓	–	✓	–	✓
division number-dependent β	–	–	✓	✓	✓	✓
time-dependent β	–	✓	–	–	✓	✓
BIC (10^4)	**3.69**	**3.44**	**3.21**	**2.57**	**3.17**	**2.60**

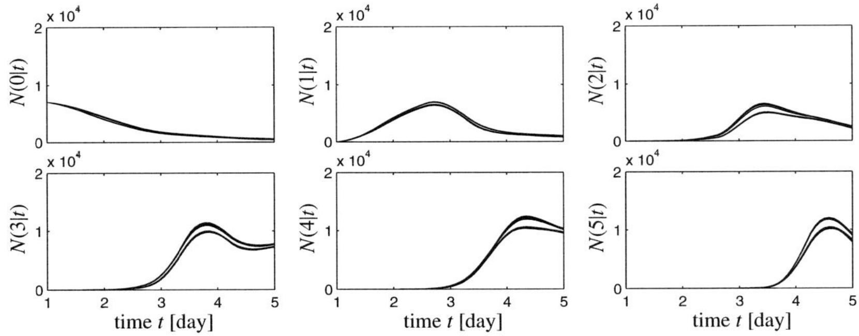

Figure 4.9: Predicted number of cells with i cell divisions, $N(i|t)$ [#], $i = 0, \ldots, 5$. The colored regions indicate the 80% ( ), 90% (■), 95% (■), and 99% (■) Bayesian confidence intervals.

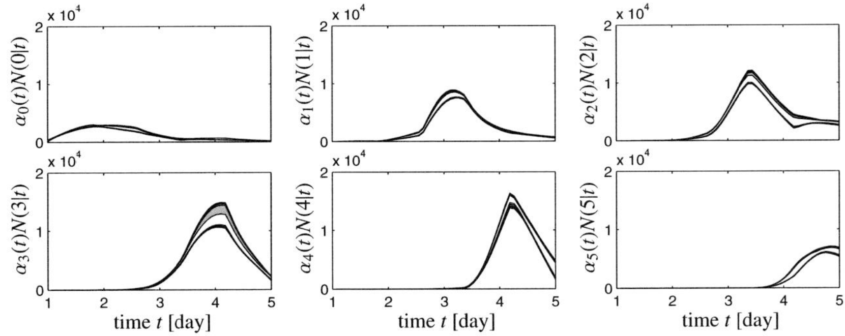

Figure 4.10: Predicted number of cells divisions per unit time in i-th subpopulations, $\alpha_i(t)N(i|t)$ [#/day], $i = 0, \ldots, 5$. The colored regions indicate the 80% ( ), 90% (■), 95% (■), and 99% (■) Bayesian confidence intervals.

4.5 Summary and discussion

In this chapter, we have proposed a division- and label-structured population model which provides a unifying framework to study proliferating cell populations which undergo symmetric cell division. This model considers both, continuous label dynamics and discrete division number-dependent effects, such as cell aging. The resulting model is a system of coupled PDEs, which, under biologically plausible assumptions, can be split up into a system of ODEs and a set of decoupled PDEs. Each PDE describes the label distribution within one particular subpopulation and the ODE model describes the number of cells per subpopulation.

We have shown that the model is a generalization of existing division-structured population models (De Boer *et al.*, 2006; Luzyanina *et al.*, 2007a) and label-structured population models (Banks *et al.*, 2011, 2010; Luzyanina *et al.*, 2009, 2007b). Both model classes can be derived from the proposed model via marginalization. In contrast to these two existing types of models, the proposed model allows to incorporate division number-dependent parameters as well as label distributions. The former one is important, as division number-dependent parameters are found in many different cell systems and are often the subject of interest, while the latter one allows the direct comparison of model predictions and data. This supersedes complex and error-prone data analysis via deconvolution or peak detection (Hawkins *et al.*, 2007; Luzyanina *et al.*, 2007a; Matera *et al.*, 2004).

Clearly, although the model provides a generalization and unification of several classes of population models, there remain models which are not covered. Examples are age-structured population models (Marciniak-Czochra *et al.*, 2009; Oldfield, 1966; Sinko & Streifer, 1967; Stiehl & Marciniak-Czochra, 2011; Trucco, 1965; von Foerster, 1959), size-structured population models (Doumic *et al.*, 2010; Sinko & Streifer, 1967), and general population balance models (Fredrickson *et al.*, 1967; Tsuchiya *et al.*, 1966). Furthermore, the size- and scar-structured population model for the asymmetrically dividing budding yeast has to be mentioned (Gyllenberg, 1986). There is quite a theory of population model construction introduced in (Diekmann *et al.*, 1998).

For the majority of these population models no analytical solutions are available. To study the dynamic properties of the models quantitatively, finite differences, finite volume, or finite elements discretization schemes are applied and the resulting ODE system is solved numerically (see, e.g., (Banks *et al.*, 2010; Luzyanina *et al.*, 2007b)). The need for numerical PDE solvers, which usually limits the state dimension to three because of the curse of dimensionality, is the main drawback of most population models. It renders the analysis complex and partially accounts for the observed focus on steady state analysis (Diekmann *et al.*, 2003, 2010; Marciniak-Czochra *et al.*, 2009; Stiehl & Marciniak-Czochra, 2011), while dynamical aspects are mostly disregarded. Furthermore, an in-depth analysis of the model and its parameters has merely been performed for one-dimensional systems.

Besides its generality, the DLSP model can also be simulated efficiently. We have proven that the solution can be approximated by a low-dimensional ODE system, employing truncation. For the truncation error we have derived an a priori bound, which can be evaluated analytically. This bound can serve to determine the minimal model order/complexity required to achieve the desired approximation quality. This renders our model better applicable in cases where many other models, e.g., (Banks *et al.*, 2011; Sinko & Streifer, 1967) come at a high computational cost.

Employing the reduced computational cost, a Bayesian parameter estimation approach has been proposed for the DLSP model. Therefore, we introduced a likelihood function which is based upon earlier work by Nüesch (2010) for discrete stochastic processes. Our likelihood function is similar to the likelihood function published recently by Thompson (2012) but assumes that cells are drawn from the population with replacement. While this is biologically not accurate, the effect of missing replacement is negligible as the size of the overall cell population is much larger than the number of drawn cells. The advantage of our likelihood function is that the different pieces of information contained in binned histogram data, namely the size of the population and the label distribution, are separated. Furthermore, we account for outliers, which increases flexibility as well as statistical robustness.

To efficiently evaluate the likelihood function, we have developed an approach using only a partial simulation of the system. Therefore, we have exploited the invariance of log-normality under the system dynamics and have derived a tight approximation of the measured fluorescence, consisting of label-induced fluorescence and autofluorescence. By intertwining model simulation and likelihood function evaluation, a significant speedup is achieved. Still, as the posterior distribution might be multimodal, efficient sampling methods are required. We suggest the usage of annealed sampling in combination with adaptive MCMC with delayed rejection for the individual updates.

To assess the properties of the modeling and estimation framework, a published dataset of T lymphocytes (Luzyanina *et al.*, 2007b) has been reanalyzed. For this dataset, we compared competing model hypotheses, including constant as well as time- and division number-dependent proliferation rates. Our model assessment using the Bayesian information criterion indicates that the rates of cell division probably depend on time and division number, while the rates of cell death merely depend on the division number. Dependencies of the proliferation rates on the label concentration – used by Luzyanina *et al.* (2009, 2007b) and by Banks *et al.* (2011, 2010, 2012) – are not required when considering division number induced heterogeneity. This is in agreement with results by (Thompson, 2012) who also compared different parameterization of the DLSP model. However, we provided the first rigorous statistical assessment using a realistic likelihood function. In previous publications merely least-squares and generalized least-squares are used as distance measures for parameter estimation. Both do not allow for a statistical interpretation for this model class.

A further disadvantage of most existing publications is that local optimization methods and tailored model for the unknown functions are used. Instead, in this work, global exploration of the parameter space via MCMC sampling is combined with highly flexible parameterization. This MCMC sampling based approach allows for a direct assessment of the parameter and prediction uncertainties. Furthermore, the flexible parameterization uncovers a delayed cell division which provides a strong indication of a minimal time between cell divisions. To consider this minimal time between subsequent cell divisions, die DLSP model can be combined with age-structured population models. While such work is in preparation (Metzger *et al.*, 2012), it is beyond the scope of this thesis.

To sum up, in this chapter we introduced the DLSP model, which accurately accounts for label dynamics and division number-dependent cell-to-cell variability. To assess the proliferation properties from labeling experiments, a Bayesian estimation methods has been introduced. To the best of our knowledge, this is the first Bayesian framework for structured population models of this type. This is probably the result of the high computational complexity associated with the evaluation of likelihood functions, which we have circumvented here.

5 Conclusion

5.1 Summary and conclusions

In this thesis, signal transduction in and proliferation of heterogenous cell populations has been investigated. For both processes, we provide novel modeling, parameter estimation, and uncertainty analysis methods. For the parameter estimation, merely single cell snapshot data are employed instead of the rarely available single cell time-lapse data used in most previous publications. As the study of cell populations is in general computationally demanding, throughout this thesis, we focused on efficient simulations and estimation procedures which make use of the respective problem structure. Therefore, we use tools from the fields of stochastic modeling, statistics, control engineering, and nonlinear optimization.

To study signal transduction in the presence of stochastic and deterministic cell-to-cell variability, an augmented Fokker-Planck equation is derived, governing the population dynamics. On the other hand, proliferating populations are described by a division- and label-structured population model. While the considered model classes are vastly different, the rigorous distinction between state dynamics and the measurement process allows for both, the derivation of partially analytical solutions and efficient simulation schemes. Especially the consideration of the superposition principle and marginalization of the state dynamics, as well as decomposition methods and invariance principles allow for simulation methods which reduce the curse of dimensionality.

The model properties also provide the basis for sophisticated Bayesian parameter estimation approaches. In Chapter 3, a parametric approach for the inference of population heterogeneity has been introduced. This parametric approach enables the decomposition of the parameter estimation into two phases. During the pre-estimation phase, an exact parametric model of the likelihood has been constructed, which can be evaluated efficiently. This parametric model of the likelihood has been used in the second phase for optimization and uncertainty analysis. Both tasks are simplified by using a parametric model of the likelihood, which renders the application of sequential convex optimization and efficient MCMC sampling feasible. This in turn allows for a rigorous analysis of the model and of the prediction uncertainties, and the detection of non-identifiability and sloppy parameters.

While the two-step procedure is rather efficient, the applicability of the approach is limited by the dimensionality of the parameter space. As an affine parameterization of the distribution is required, common ansatz function choices are probably restricted to three or four parameters. To push this limit, recursive refinement methods may be used. However, also these methods have limits. In addition, it is known that, while some parameters may vary among cells, e.g., protein synthesis rates, others are identical in all cells, e.g., affinities. The current approach does not allow for a simultaneous estimation of the distribution of the cell-specific parameters and the value of the shared parameter. Despite these shortcomings, the proposed scheme extends previous results and the example in Section 3.4 illustrates its benefits in case of a small number of unknown parameter distributions.

Analoguously to parameter estimation for the augmented Fokker-Planck equation, we have proposed in Chapter 4 a parameter estimation scheme for the division- and label-structured population model which exploits the model properties. While no parametric model of the likelihood function could be derived, we circumvent the need to simulate the overall PDE model to evaluate the likelihood function. Instead, the likelihood is assessed by exploiting the decomposition and invariance principles, which yields an approach only requiring the simulation of a low-dimensional ODE model. Accordingly, the likelihood function can be evaluated extremely efficiently, allowing for the investigation of the posterior distribution. This yields global Bayesian confidence intervals for parameters, unknown functions (which are parameterized), as well as model predictions.

Despite the computationally efficient evaluation of the likelihood function, the remaining parameter estimation problem is still challenging. It is nonlinear and the posterior distribution may possess different modes, rendering the application of global optimization and uncertainty analysis schemes essential. While sophisticated MCMC sampling algorithms, employing annealing, adaptation and delayed rejection, can partially overcome this problem, for increasing problem dimensionality the convergence rate decrease. Therefore, it is crucial to find flexible parameterizations of the unknown functions which possess a small number of parameters. As shown in the example in Section 4.4, flexibility is crucial to avoid misleading interpretations and to allow for an unbiased analysis.

Regarding methodology and theoretical concepts: The methods used for the analysis of the individual model classes are similar and might be used to study other population models as well. However, their application and the final implementation is highly problem-specific. As for single cell models, different formulations of the estimation problem are used depending on the considered model classes, e.g., Markov jump process, chemical Langevin equations, and reaction rate equations. Beyond this, the formulation strongly depends on the type of available measurement data.

In summary, the methods proposed in Chapter 3 and 4 of this thesis constitute a significant progress towards uncertainty-aware computational modeling of heterogeneous cell populations. A variety of the developed ideas and concepts are rather general and might be applicable to other classes of population models, data types, and biological questions.

5.2 Outlook

In the following, we outline open research problems and opportunities which arise from the results of this thesis.

5.2.1 Parameter estimation from multiple data types

Throughout this work, we consider single cell snapshot data. This data type is known to provides reliable population statistics, but the information about the individual cells is limited, as cells are not tracked over time. This is in contrast to single cell time-lapse data which provide more information about the individual cells, but worse population statistics, as the number of measured cells is generally smaller. Hence, single cell snapshot data and single cell time-lapse data provide, to a certain degree, complementary information. To reduce parameter uncertainties and non-identifiability, methods for simultaneous parameter estimation from different data sources have to be developed.

The development of such “multiple data type parameter estimation methods” are the logical consequence of the increased availability of different data types due to technology advances. Nowadays, it becomes more and more common to work in large research consortia with a variety of experimental groups with different experimental expertise and capable of providing different data types. A holistic understanding of systems requires the integration of all these pieces of information. Nevertheless, appropriate methods are missing, mainly due to the lack of computationally efficient formulations.

The efficient parameter estimation methods for single cell snapshot data proposed in this thesis render multiple data type parameter estimation methods feasible. In particular, the proposed Bayesian methods for single cell snapshot data can be integrated with available methods for single cell time-lapse data (Koeppl *et al.*, 2012), both for signal transduction and proliferation processes. The simultaneous handling of these data types requires models on the single cell and the population level. While the single cell snapshot data, $\mathcal{D}_{ss}$, might be compared to the predictions of a population, the observed single cell time-lapse data, $\mathcal{D}_{tl}$, are compared to the prediction of single cell models with varying parameters. This would yield an overall likelihood function,

$$\mathbb{P}(\mathcal{D}_{ss}, \mathcal{D}_{tl} | p_\theta, \{\theta^i\}_{i=1}^{d_{\mathcal{D}_{tl}}}) = \mathbb{P}(\mathcal{D}_{ss}|p_\theta) \left(\prod_{i=1}^{d_{\mathcal{D}_{tl}}} \mathbb{P}(\mathcal{D}_{tl}^i|\theta^i) p_\theta(\theta^i) \right), \tag{5.1}$$

in which $\mathcal{D}_{tl}^i$ is the individual time-lapse trajectory. By employing an affine parametrization of $p_\theta(\theta)$ and the previously presented two-step procedure, the computational complexity of evaluating $\mathbb{P}(\mathcal{D}_{ss}|p_\theta)$ can be reduced. Hence, merely the simulation of a medium number of single cell models remains to assess the likelihood for a particular parameter distribution. Despite this, the problem remains challenging, as the parameter vector is high-dimensional and includes the weighting of the ansatz functions, as well as a parameter for each single cell for which single time-lapse data are available – this is similar to work by Koeppl *et al.* (2012).

In addition, the ansatz function based approach does not allow us to distinguish between parameters which vary among cells, such as the abundance of transcription factors, and parameters which are similar for all cells, e.g., protein-protein affinities. The incorporation of these different parameter classes would be crucial, as discussed above.

5.2.2 Reduced order modeling for parameter estimation

In Chapter 3 and 4, we presented two methods to tackle the computational complexity of parameter estimation for cell population models. In our opinion, these methods represent two extremes. While the computation of a parametric form of the likelihood changes the estimation problem completely, the intertwining of simulation and likelihood function evaluation also results in a tremendous speedup but conserves the structure of the problem.

Apparently, there are many problem formulations in between these two extremes. We argue that especially the combination of parameter estimation and parametric model reduction is promising. The idea is again to split up the parameter estimation and uncertainty analysis into two steps (Benner, 2009; Hasenauer *et al.*, 2012b). In the first step, parametric model reduction is employed to derive a parametric reduced order model. This parametric reduced order model depends on the original parameters, but can be simulated more efficiently. During the parameter estimation, this reduced model is used to evaluate the likelihood function for different parameter values. If the reduced order model provides a good approximation of the output of the population model, this scheme provides a reasonable approximation to the original parameter estimation, while resulting in a reduced computational burden.

The main challenges of this approach are the derivation of a parametric reduced order model and checking/accounting for approximation errors. While common projection based approaches (Antoulas, 2005a,b; Bai, 2002; Freund, 2003; Haasdonk & Ohlberger, 2011) may be employed to study high-dimensional linear systems, for nonlinear systems, like the (augmented) Fokker-Planck equation, nonlinear methods are required. Such methods are in general only applicable for deterministic single cell models (see, e.g., (Löhning *et al.*, 2011a,b) and references therein). To approximate the statistics of stochastic models, moment-based methods may be employed, as discussed by (Zechner *et al.*, 2012).

Beyond the derivation of parametric reduced order models, in some situations also a direct approximation of the likelihood function may be possible. Therefore, e.g., polynomial chaos expansions (Oladyshkin *et al.*, 2011; Villadsen & Michelsen, 1978) may be employed. As the parameter estimation is a sequential process, the derivation of the reduced model might be integrated with the search itself.

5.2.3 Analysis of population models

Although cell population models have become increasingly common, tools for their analysis, besides simple simulation, are still missing. This problem originates from the complexity of population models. Of particular practical interest are methods to assess the subpopulation structure (Song *et al.*, 2010) and to determine the main sources of cell-to-cell variability (Snijder & Pelkmans, 2011; Hasenauer *et al.*, 2011a). While many parameters may alter the quantitative behavior of cells, in many situations only few will influence decision making. This has been shown, e.g., for apoptosis induction (see (Spencer *et al.*, 2009)), where under a variety of conditions only few factors influence the life-death decision.

Most methods used to detect these decision markers which determine the phenotype of cells are limited and merely employ simulation. Apparently, to enable a sound model-based assessment of the subpopulation structure, sophisticated classification tools are required. We presented approaches tackling this problem (Hasenauer *et al.*, 2011a, 2012a), which combine an intuitive assessment of the high-dimensional dependencies via parallel-coordinate plots (Inselberg & Dimsdale, 1990; McDonnell & Mueller, 2008) with a quantification of dependencies via nonlinear support vector machines (Ivanciuc, 2007; Schölkopf *et al.*, 1997; Smola & Schölkopf, 2004). However, these approaches do not account for model uncertainties. Furthermore, an assessment of the parameter dependent population structure, similar to the bifurcation analysis presented by Song *et al.* (2010), is so far not feasible.

To approach these problems, analysis schemes for PDE models have to be developed, potentially employing the aforementioned model properties. In particular, the generalization of the concept of separatrices, used in some recent studies (Aldridge *et al.*, 2011), for the case of stochastic models would be essential. It could allow for the efficient computation of cell fate probabilities, which are the implicit design objective in many studies.

In conclusion, the presented modeling and estimation framework opens up many research directions. Possibilities range from more predictive models over new analysis tools towards the integration of multiple data sources and types. These are crucial steps towards a holistic understanding of biological processes.

Appendix

A. Proof of analytical solution of PDE (4.26)

To determine the solution of the PDE (4.26) the method of characteristics (Evans, 1998) is employed, which is possible as (4.26) is linear. The characteristics of (4.26) are defined by the ODEs

$$\frac{dx}{d\tau} = -k(t)x, \quad \frac{dt}{d\tau} = 1, \quad \frac{dp_i}{d\tau} = k(t)p_i, \tag{A.1}$$

with $x(0) = x_0$, $t(0) = 0$, and $p_i(x_0) = \gamma^i p_0(\gamma^i x_0)$. This system of ODEs has the solution

$$x(\tau) = x_0 e^{-\int_0^\tau k(\tilde{\tau})d\tilde{\tau}}, \quad t(\tau) = \tau, \quad p_i(\tau) = \gamma^i e^{\int_0^\tau k(\tilde{\tau})d\tilde{\tau}} p_0(\gamma^i x_0). \tag{A.2}$$

By substitution we obtain

$$p(x|i,t) = \gamma^i e^{\int_0^t k(\tau)d\tau} p_0(\gamma^i e^{\int_0^t k(\tau)d\tau} x) \tag{A.3}$$

as solution for (4.26).

B. Proof of Lemma 4.5: Solution of ODE system

In this section we prove by mathematical induction that the ODE system

$$\begin{aligned} i = 0: \ & \frac{dN}{dt}(0|t) = -(\breve{\alpha} + \beta)\, N(0|t), \\ \forall i \geq 1: \ & \frac{dN}{dt}(i|t) = -(\breve{\alpha} + \beta)\, N(i|t) + 2\hat{\alpha} N(i-1|t) \end{aligned} \tag{B.1}$$

with initial conditions $N(0|0) = N_0$ and $\forall i \geq 1 : N(i|0) = 0$, has for $\hat{\alpha}, \breve{\alpha} \geq 0$ and $\beta > 0$ the solution:

$$N(i|t) = \frac{(2\hat{\alpha}t)^i}{i!} e^{-(\breve{\alpha}+\beta)t} N_0. \tag{B.2}$$

Thereby, (B.1) is a generalization of (4.36).

It is trivial to verify that $N(0|t)$ and $N(1|t)$ are the solutions of (B.1) for $i = 0$ and $i = 1$, respectively. Hence, only the problem of proving that $N(k+1|t)$ is the solution of (E.1) for $i = k+1$ given $N(k|t)$ remains. To show this, note that

$$(\text{B.2}) \quad \circ\!\!-\!\!\bullet \quad \forall i \in \mathbb{N}_0 : \mathcal{N}(i|s) = \frac{(2\hat{\alpha})^i}{(s + \breve{\alpha} + \beta)^{i+1}} N_0, \tag{B.3}$$

in which $\mathcal{N}(i|s)$ is the Laplace transform of $N(i|t)$. Given this

$$\begin{aligned} &\frac{dN}{dt}(k+1|t) = -(\check{\alpha}+\beta)\,N(k+1|t) + 2\hat{\alpha}N(k|t) \\ \circ\!\!-\!\!\bullet\quad & s\mathcal{N}(k+1|s) = -(\check{\alpha}+\beta)\,\mathcal{N}(k+1|s) + 2\hat{\alpha}\mathcal{N}(k|s) \\ \Leftrightarrow\quad & \mathcal{N}(k+1|s) = \frac{2\hat{\alpha}}{s+\check{\alpha}+\beta}\mathcal{N}(k|s). \end{aligned} \tag{B.4}$$

Substitution of $\mathcal{N}(k|s)$ now yields,

$$\mathcal{N}(k+1|s) = \frac{(2\hat{\alpha})^{k+1}}{(s+\check{\alpha}+\beta)^{k+2}}N_0 \tag{B.5}$$

which by applying the inverse Laplace transformation concludes the mathematical induction and proves Lemma 4.5.

Remark .1. *Note that for* $\check{\alpha} = \hat{\alpha} = \alpha$, (B.2) *simplifies to* (4.36). *While for* $\hat{\alpha} = \alpha_{\text{sup}}$, $\check{\alpha} = \alpha_{\text{inf}}$, $\beta = \beta_{\text{inf}}$, $N(i|t) = B(i|t)$, *and* $N_0 = B_0$, *we obtain the bounding system* (E.1) *and its solution.*

C. Proof of Lemma 4.6: Solution of ODE system

In this section we prove that if

- $\forall i : \alpha_i(t) = \alpha_i \;\wedge\; \beta_i(t) = \beta_i$ and
- $\forall i, j \in \mathbb{N}_0, i \neq j : \alpha_i + \beta_i \neq \alpha_j + \beta_j$

the solution of (4.25) is

$$\begin{aligned} i = 0 &: N(0|t) = e^{-(\alpha_0+\beta_0)t}N_0 \\ \forall i \geq 1 &: N(i|t) = 2^i\left(\prod_{j=1}^{i}\alpha_{j-1}\right)D_i(t)N_0 \end{aligned} \tag{C.1}$$

in which

$$D_i(t) = \sum_{j=0}^{i}\left[\left(\prod_{\substack{k=0\\k\neq j}}^{i}((\alpha_k+\beta_k)-(\alpha_j+\beta_j))\right)^{-1} e^{-(\alpha_j+\beta_j)t}\right].$$

It is not difficult to verify that $N(0|t)$ and $N(1|t)$ are the solutions of (4.25) for $i = 0$ and $i = 1$, respectively. Hence, only the problem of proving that $N(k+1|t)$ is the solution of (E.1) for $i = k+1$ given $N(k|t)$ remains. To show this, note that for

$$\text{(C.1)}\quad \circ\!\!-\!\!\bullet\quad \forall i \in \mathbb{N}_0 : \mathcal{N}(i|s) = 2^i\frac{\prod_{j=1}^{i}\alpha_{j-1}}{\prod_{j=0}^{i}(s+\alpha_j+\beta_j)}N_0, \tag{C.2}$$

in which $\mathcal{N}(i|s)$ is the Laplace transform of $N(i|t)$. The proof of this relation is provided in Appendix D.

Given (C.2) it follows that

$$
\begin{aligned}
& \frac{dN}{dt}(k+1|t) = -(\alpha_{k+1}+\beta_{k+1})\,N(k+1|t) + 2\alpha_k N(k|t) \\
\circ\!\!-\!\!\bullet\quad & s\mathcal{N}(k+1|s) = -(\alpha_{k+1}+\beta_{k+1})\,\mathcal{N}(k+1|s) + 2\alpha_k \mathcal{N}(k|s) \\
\Leftrightarrow\quad & \mathcal{N}(k+1|s) = \frac{2\alpha_k}{s+\alpha_{k+1}+\beta_{k+1}}\mathcal{N}(k|s).
\end{aligned}
\tag{C.3}
$$

Substitution of $\mathcal{N}(k|s)$ now yields,

$$
\mathcal{N}(k+1|s) = 2^i \frac{\prod_{j=1}^{k+1} \alpha_{j-1}}{\prod_{j=0}^{k+1}(s+\alpha_j+\beta_j)} N_0 \tag{C.4}
$$

which by applying the inverse Laplace transformation concludes the mathematical induction and proves (B.2).

D. Derivation of Laplace transform $\mathcal{N}(i|s)$

To derive $\mathcal{N}(i|s)$ defined in (C.2), we study the partial fraction of

$$
\mathcal{N}(i|s) = 2^i \frac{\prod_{j=1}^{i} \alpha_{j-1}}{\prod_{j=0}^{i}(s+\alpha_j+\beta_j)} N_0. \tag{D.1}
$$

As under the prerequisite $\forall i, j \in \mathbb{N}_0$ with $i \neq j : \alpha_i + \beta_i \neq \alpha_j + \beta_j$ all poles are distinct, the partial fraction can be written as

$$
\mathcal{N}(i|s) = 2^i \left(\prod_{j=1}^{i} \alpha_{j-1} \right) \left(\sum_{k=0}^{i} \frac{c_k}{(s+\alpha_k+\beta_k)} \right) N_0. \tag{D.2}
$$

To determine the coefficients c_k, we consider the equality constraint

$$
\begin{aligned}
\frac{1}{\prod_{j=0}^{i}(s+\alpha_j+\beta_j)} &= \sum_{k=0}^{i} \frac{c_k}{(s+\alpha_k+\beta_k)} \\
\Leftrightarrow\qquad 1 &= \sum_{k=0}^{i} c_k \prod_{\substack{j=1 \\ j\neq k}}^{i} (s+\alpha_j+\beta_j).
\end{aligned}
\tag{D.3}
$$

As this equality constraint has to hold for all s, it must be satisfied for $s = -(\alpha_k + \beta_k)$, yielding

$$
c_k = \left(\prod_{\substack{j=1 \\ j\neq k}}^{i} ((\alpha_j+\beta_j) - (\alpha_k+\beta_k)) \right)^{-1}. \tag{D.4}
$$

Given the values for c_k one can easily verify (C.2) by plugging in the c_k's into (D.2). Obviously, the proposed procedure can also be inverted, which concludes the derivation of (C.2).

E. Proof of Theorem 4.7: Convergence

To prove Theorem 4.7, the comparison theorem for series (Knopp, 1964) is applied. Therefore, we define the bounding system

$$\begin{aligned} i = 0: \quad & \frac{dB}{dt}(0|t) = -(\alpha_{\inf} + \beta_{\inf})\, B(0|t), \\ \forall i \geq 1: \quad & \frac{dB}{dt}(i|t) = -(\alpha_{\inf} + \beta_{\inf})\, B(i|t) + 2\alpha_{\sup} B(i-1|t) \end{aligned} \tag{E.1}$$

with initial conditions

$$i = 0: B(0|0) = N_0, \quad \forall i \geq 1: B(i|0) = 0$$

and $\alpha_{\inf}$, $\alpha_{\sup}$, and $\beta_{\inf}$ as in Theorem 4.7. Due to the simple structure of (E.1), we can compute the analytical solution

$$B(i|t) = \frac{(2\alpha_{\sup} t)^i}{i!} e^{-(\alpha_{\inf}+\beta_{\inf})t} N_0, \tag{E.2}$$

whose derivation can be found in Appendix B.

The bounding system (E.1) is obtained from (4.25) by reducing the outflows out of and increasing the inflows into the individual subpopulations. Intuitively, as the initial conditions of (E.1) and (4.25) are identical and the right hand side of (E.1) is for every $t \in [0, T]$ greater or equal than the right hand side of (4.25), it follows that B_i is an upper bound for N_i,

$$\forall t \in [0, T], i: \quad B(i|t) \geq N(i|t). \tag{E.3}$$

This can be proven rigorously by applying Müller's theorem (Müller, 1927), as shown in (Kieffer & Walter, 2011) for another system.

Given (E.2) and (E.3) one can prove the convergence of $\sum_{i \in \mathbb{N}_0} n(x, i|t)$. To take into account that a distributed process is considered ($x \geq 0$), we study the maximum over x and define $B_i(t) := B(i|t)\gamma^i e^{kt} p_0^{\sup} = \frac{(2\alpha_{\sup}\gamma)^i}{i!} t^i e^{-(\alpha_{\inf}+\beta_{\inf})t} e^{kt} n_0^{\sup}$ with $p_0^{\sup} := \sup_{x \in \mathbb{R}_+} p_0(x)$ and $n_0^{\sup} := N_0 p_0^{\sup}$. Thus, $B_i(t)$ is a point-wise upper bound of $n(x, i|t)$. For this definition of $B_i(t)$ it holds that

(i) $\forall i, t, x \geq 0: \; 0 \leq N_i(t, x) \leq B_i(t) \; \forall i$, and

(ii) the series

$$\sum_{i=0}^{\infty} B_i(t) = \left(\sum_{i=0}^{\infty} \frac{(2\alpha_{\sup}\gamma t)^i}{i!} \right) e^{-(\alpha_{\inf}+\beta_{\inf})t} e^{kt} n_0^{\sup} \tag{E.4}$$

is convergent for every finite t.

The latter one holds true as the series is simply the Taylor expansion of the exponential $e^{2\alpha_{\sup}\gamma t}$. Under conditions (i) and (ii) it follows from the comparison theorem for series (Knopp, 1964) that the series $\sum_{i \in \mathbb{N}_0} N(i|t)$ is convergent in i for every $t \in [0, T]$ and for every $x \geq 0$. This concludes the proof.

F. Proof of Theorem 4.8: Truncation error

To prove Theorem 4.8, note that

$$\begin{aligned} \|n(x|T) - \hat{n}_S(x|T)\|_1 &= \left\| \sum_{i=S}^{\infty} N(i|T) p(x|i,T) \right\|_1 \\ &= \sum_{i=S}^{\infty} N(i|T) \int_{\mathbb{R}_+} p(x|i,T) dx \\ &= \sum_{i=S}^{\infty} N(i|T), \end{aligned} \tag{F.1}$$

in which the individual lines follow from the approximation methods (4.38), the fact that all quantities are positive, and the definition of the normalized label intensity (4.26) which has unity integral for all times $T \geq 0$. The remaining term in the following is successively upper bounded, for which we employ the bounding system (E.1). As shown in Appendix E, it holds that $N(i|t) \leq B(i|t)$ which yields

$$\sum_{i=S}^{\infty} N(i|T) \leq \sum_{i=S}^{\infty} B(i|T) = \sum_{i=S}^{\infty} \frac{(2\alpha_{\sup} T)^i}{i!} e^{-(\alpha_{\inf}+\beta_{\inf})T} N_0. \tag{F.2}$$

By completion of the sum, this can be written as

$$\sum_{i=S}^{\infty} N(i|T) \leq \left(e^{2\alpha_{\sup} T} - \sum_{i=0}^{S-1} \frac{(2\alpha_{\sup} T)^i}{i!} \right) e^{-(\alpha_{\inf}+\beta_{\inf})T} N_0. \tag{F.3}$$

Thus, by exploiting that $\|n(x|0)\|_1 = N_0$, one obtains (4.40), which concludes the proof.

G. Proof that the solution of LSP can be constructed from DLSP

To prove that the DLSP provides the solution to the LSP, $n^{\mathrm{LSP}}(x|t) = n(x|t)$, we show that $n(x|t) = \sum_{i\in\mathbb{N}_0} N(i|t)p(x|i,t)$ solves (4.46). Therefore, $n(x|t)$ is inserted in the left hand side $(*)$ of (4.46), yielding

$$\begin{aligned} (*) &= \frac{\partial}{\partial t}\left(\sum_{i\in\mathbb{N}_0} N(i|t)p(x|i,t) \right) - k\frac{\partial}{\partial x}\left(x \sum_{i\in\mathbb{N}_0} N(i|t)p(x|i,t) \right) \\ &= \sum_{i\in\mathbb{N}_0} \left(\frac{dN}{dt}(i|t)p(x|i,t) + \underbrace{N(i|t)\left(\frac{\partial p(x|i,t)}{\partial t} - k\frac{\partial (xp(x|i,t))}{\partial x} \right)}_{=0 \text{ (with (4.26))}} \right). \end{aligned}$$

In here, $dN(i|t)/dt$ is substituted with (4.25), resulting in

$$\begin{aligned} (*) &= \sum_{i\in\mathbb{N}_0} (-(\alpha(t)+\beta(t))N(i|t)p(x|i,t)) + \sum_{i\in\mathbb{N}} 2\alpha(t)N(i-1|t)\underbrace{p(x|i,t)}_{=\gamma p(\gamma x|i-1,t)} \\ &= -(\alpha(t)+\beta(t))\sum_{i\in\mathbb{N}_0} N(i|t)p(x|i,t) + 2\gamma\alpha(t)\sum_{i\in\mathbb{N}_0} N(i|t)n_i(t,\gamma x). \end{aligned}$$

This is equivalent to the result if $n(x|t)$ is inserted in the right hand side $(*)$ of (4.46). Hence, $n(x|t) = \sum_{i\in\mathbb{N}_0} N(i|t)p(x|i,t)$ fulfills (4.46) which concludes the proof.

H. Proof that the PDE (4.26) conserves log-normal distributions

To prove that the PDE (4.26) conserves log-normal distributions, we use its analytical solution (4.35) and consider $p_0(x) = \log\mathcal{N}(x|\mu_0, \sigma_0^2)$. This yields the solution

$$p(x|i,t) = \gamma^i e^{\int_0^t k(\tau)d\tau} \log\mathcal{N}(\gamma^i e^{\int_0^t k(\tau)d\tau} x|\mu_0, \sigma_0^2). \tag{H.1}$$

Employing the definition of the log-normal distribution, this equation becomes

$$p(x|i,t) = \gamma^i e^{\int_0^t k(\tau)d\tau} \frac{1}{\sqrt{2\pi}\sigma_0 \left(\gamma^i e^{\int_0^t k(\tau)d\tau} x\right)} e^{-\frac{1}{2}\left(\frac{\log\left(\gamma^i e^{\int_0^t k(\tau)d\tau} x\right)-\mu_0}{\sigma_0}\right)^2} \tag{H.2}$$

$$= \frac{1}{\sqrt{2\pi}\sigma_0 x} e^{-\frac{1}{2}\left(\frac{\log x - \left(-i\log\gamma - \int_0^t k(\tau)d\tau + \mu_0\right)}{\sigma_0}\right)^2}, \tag{H.3}$$

for $x > 0$, which can be restated as

$$p(x|i,t) = \log\mathcal{N}(x|\mu_i(t), \sigma_0^2), \tag{H.4}$$

in which $\mu_i(t) = -i\log\gamma - \int_0^t k(\tau)d\tau + \mu_0$. As this equation also holds for $x \leq 0$, it follows that the log-normal distribution is conserved and merely the parameter μ is time dependent. Employing the superposition principle, this statement can be directly extended for sums of log-normal distributions, which concludes the proof.

Bibliography

Al-Banna, M. K., Kelman, A. W., & Whiting, B. (1990). Experimental design and efficient parameter estimation in population pharmacokinetics. *J. Pharmacokin. Biopharm.*, *18*(4), 347–360.

Albeck, J. G., Burke, J. M., Aldridge, B. B., Zhang, M., Lauffenburger, D. A., & Sorger, P. K. (2008a). Quantitative analysis of pathways controlling extrinsic apoptosis in single cells. *Mol. Cell*, *30*(1), 11–25.

Albeck, J. G., Burke, J. M., Spencer, S. L., Lauffenburger, D. A., & Sorger, P. K. (2008b). Modeling a snap-action, variable-delay switch controlling extrinsic cell death. *PLoS Biol.*, *6*(12), 2831–2852.

Aldridge, B. B., Gaudet, S., Lauffenburger, D. A., & Sorger, P. K. (2011). Lyapunov exponents and phase diagrams reveal multi-factorial control over TRAIL-induced apoptosis. *Mol. Syst. Biol.*, *7*, 553.

Anderson, A. R. A. & Quaranta, V. (2008). Integrative mathematical oncology. *Nat. Rev. Cancer*, *8*(3), 227–234.

Andres, C., Hasenauer, J., Allgöwer, F., & Hucho, T. (2012). Threshold-free population analysis identifies larger DRG neurons to respond stronger to NGF stimulation. *PLoS ONE*, *7*(3), e34257.

Andres, C., Meyer, S., Dina, O. A., Levine, J. D., & Hucho, T. (2010). Quantitative automated microscopy (QuAM) elucidates growth factor specific signalling in pain sensitization. *Molecular Pain*, *6*(98), 1–16.

Angeli, D., Ferrell Jr., J. E., & Sontag, E. D. (2004). Detection of multistability, bifurcations, and hysteresis in a large class of biological positive-feedback systems. *Proc. Nati. Acad. Sci. U S A*, *101*(7), 1822–1827.

Antoulas, A. C. (2005a). *Approximation of large-scale dynamical systems*. Advances in design and control. Philadelphia: SIAM.

Antoulas, A. C. (2005b). An overview of approximation methods for large-scale dynamical systems. *Annu. Rev. Control*, *29*(2), 181–190.

Apgar, J. F., Witmer, D. K., White, F. M., & Tidor, B. (2010). Sloppy models, parameter uncertainty, and the role of experimental design. *Mol. BioSyst.*, *6*(10), 1890–1900.

Audoly, S., Bellu, G., D'Angiò, L., Saccomani, M. P., & Cobelli, C. (2001). Global identifiability of nonlinear models of biological systems. *IEEE Trans. Biomed. Eng.*, *48*(1), 55–65.

Avery, S. V. (2006). Microbial cell individuality and the underlying sources of heterogeneity. *Nat. Rev. Microbiol.*, *4*, 577–587.

Bai, Z. (2002). Krylov subspace techniques for reduced-order modeling of large-scale dynamical systems. *Appl. Numer. Math.*, *43*(1–2), 9–44.

Balsa-Canto, E., Alonso, A. A., & Banga, J. R. (2010). An iterative identification procedure for dynamic modeling of biochemical networks. *BMC Syst. Biol.*, *4*(11).

Balsa-Canto, E., Peifer, M., Banga, J. R., Timmer, J., & Fleck, C. (2008). Hybrid optimization method with general switching strategy for parameter estimation. *BMC Syst. Biol.*, *2*(26).

Banga, J. R. (2008). Optimization in computational systems biology. *BMC Syst. Biol.*, *2*(47).

Banks, H. T., Sutton, K. L., Thompson, W. C., Bocharov, G., Doumic, M., Schenkel, T., Argilaguet, J., Giest, S., Peligero, C., & Meyerhans, A. (2011). A new model for the estimation of cell proliferation dynamics using CFSE data. *J. Immunological Methods*, *373*(1–2), 143–160.

Banks, H. T., Suttona, K. L., Thompson, W. C., Bocharov, G., Roose, D., Schenkel, T., & Meyerhans, A. (2010). Estimation of cell proliferation dynamics using CFSE data. *Bull. Math. Biol.*, *73*(1), 116–150.

Banks, H. T., Thompson, W. C., Peligero, C., Giest, S., Argilaguet, J., & Meyerhans, A. (2012). A division-dependent compartmental model for computing cell numbers in CFSE-based lymphocyte proliferation assays. Technical Report CRSC-TR12-03, Center for Research in Scientific Computation, North Carolina State University, North Carolina, USA.

Beaulieu, N. C. (2004). Highly accurate simple closed-form approximations to lognormal sum distributions and densities. *IEEE Commun. Lett.*, *8*(12), 709–711.

Beaumont, M. A., Zhang, W., & Balding, D. J. (2002). Approximate Bayesian computation in population genetics. *Genetics*, *162*(4), 2025–2035.

Benner, P. (2009). System-theoretic methods for model reduction of large-scale systems: Simulation, control, and inverse problems. In *Proc. of the Int. Conf. on Math. Modelling, Vienna, Austria*. 126–145.

Berg, O. G. (1978). A model for the statistical fluctuations of proteins numbers in a microbial population. *J. Theor. Biol.*, *71*(4), 587–603.

Biegler, L. T. (2007). An overview of simultaneous strategies for dynamic optimization. *Chem. Eng. Process.*, *46*(11), 1043–1053.

Bird, J. J., Brown, D. R., Mullen, A. C., Moskowitz, N. H., Mahowald, M. A., Sider, J. R., Gajewski, T. F., Wang, C.-R., & Reiner, S. L. (1998). Helper T cell differentiation is controlled by the cell cycle. *Immunity*, *9*(2), 229–237.

Bock, H. H. G. & Plitt, K. J. (1984). A multiple shooting algorithm for direct solution of optimal control problems. In *Proc. of the 9th IFAC World Congress, Budapest, Hungary*. 242–247.

Botev, Z. I., Grotowiski, J. F., & Kroese, D. P. (2010). Kernel density estimation via diffusion. *Ann. Statist.*, *38*(5), 2916–2957.

Boyd, S. & Vandenberghe, L. (2004). *Convex Optimisation*. Cambridge University Press, UK.

Brännmark, C., Palmer, R., Glad, S. T., Cedersund, G., & Strålfors, P. (2010). Mass and information feedbacks through receptor endocytosis govern insulin signaling as revealed using a parameter-free modeling framework. *J. Biol. Chem.*, *285*(26), 20171–20179.

Brooks, S. P. & Roberts, G. O. (1998). Assessing convergence of Markov chain Monte Carlo algorithms. *Stat. Comp.*, *8*(4), 319–335.

Brown, K. S., Hill, C. C., Calero, G. A., Myers, C. R., Lee, K. H., Sethna, J. P., & Cerione, R. A. (2004). The statistical mechanics of complex signaling networks: Nerve growth factor signaling. *Phys. Biol.*, *1*, 184–195.

Brown, M. R., Summers, H. D., Rees, P., Smith, P. J., Chappell, S. C., & Errington, R. J. (2010). Flow-based cytometric analysis of cell cycle via simulated cell populations. *PLoS Comput. Biol.*, *6*(4), e1000741.

Buckwar, E. & Winkler, R. (2004). On two-step schemes for SDEs with small noise. In *Proc. Appl. Math. Mech. (PAMM)*. Wiley-VCH, Weinheim, vol. 4, 15–18.

Buckwar, E. & Winkler, R. (2006). Multistep methods for SDEs and their application to problems with small noise. *SIAM J. Numer. Anal.*, *44*(2), 779–803.

Busetto, A. G. & Buhmann, J. M. (2009). Structure identification by optimized interventions. In *Proc. of the 12th Int. Conf. on Artif. Intell. and Stat.*. 57–64.

Busetto, A. G., Ong, C. S., & Buhmann, J. M. (2009). Optimized expected information gain for nonlinear dynamical systems. In *Proc. of the 26th Annu. Int. Conf. on Mach. Learn., ICML 2009*. ACM, vol. 382 of *ACM Int. Conf. Proc. Ser.*, 13.

Buske, P., Galle, J., Barker, N., Aust, G., Clevers, H., & Loeffler, M. (2011). A comprehensive model of the spatio-temporal stem cell and tissue organisation in the intestinal crypt. *PLoS Comput. Biol.*, *7*(1), e1001045.

Carpenter, J. & Bithell, J. (2000). Bootstrap confidence intervals: when, which, what? A practical guide for medical statisticians. *Statistics in medicine*, *19*, 1141–1164.

Chaves, M., Eissing, T., & Allgöwer, F. (2008). Bistable biological systems: A characterization through local compact input-to-state stability. *IEEE Trans. Autom. Control*, *53*, 87–100.

Chen, M.-H. & Shao, Q.-M. (1999). Monte Carlo estimation of Bayesian credible and HPD intervals. *J. Comput. Graphical Statist.*, *8*(1), 69–92.

Chen, W. W., Niepel, M., & Sorger, P. K. (2010). Classic and contemporary approaches to modeling biochemical reactions. *Genes. Dev.*, *24*(17), 1861–1875.

Cheong, R., Hoffmann, A., & Levchenko, A. (2008). Understanding nf-κb signaling via mathematical modeling. *Mol. Syst. Biol.*, *4*(192), 1–11.

Chib, S. & Greenberg, E. (1995). Understanding the Metropolis-Hastings algorithm. *Am. Stat.*, *49*(4), 327—335.

Cobelli, C. & DiStefano, J. J. (1980). Parameter and structural identifiability concepts and ambiguities: a critical review and analysis. *Am. J. Physiol. Regul. Integr. Comp. Physiol.*, *239*(1), 7–24.

Dargatz, C. (2010). *Bayesian inference for diffusion processes with applications in life sciences*. Ph.D. thesis, Ludwig–Maximilians–Universität München, Germany.

De Boer, R. J., Ganusov, V. V., Milutinoviò, D., Hodgkin, P. D., & Perelson, A. S. (2006). Estimating lymphocyte division and death rates from CFSE data. *Bull. Math. Biol.*, *68*(5), 1011–1031.

De Boor, C. (1978). *A practical guide to splines*, vol. 27 of *Applied Mathematical Sciences*. New York: Springer Verlag.

Deenick, E. K., Gett, A. V., & Hodgkin, P. D. (2003). Stochastic model of T cell proliferation: A calculus revealing IL-2 regulation of precursor frequencies, cell cycle time, and survival. *J. Immunol.*, *170*(10), 4963–4972.

DiCiccio, T. J. & Efron, B. (1996). Bootstrap confidence intervals. *Statist. Sci.*, *11*(3), 189–228.

Diekmann, O., Gyllenberg, M., & Metz, J. A. J. (2003). Steady state analysis of structured population models. *Theor. Population Biol.*, *63*(4), 309–338.

Diekmann, O., Gyllenberg, M., Metz, J. A. J., Nakaoka, S., & de Roos, A. M. (2010). Daphnia revisited: local stability and bifurcation theory for physiologically structured population models explained by way of an example. *J. Math. Biol.*, *62*(2), 277–318.

Diekmann, O., Gyllenberg, M., Metz, J. A. J., & Thieme, H. R. (1998). On the formulation and analysis of general deterministic structured population models. I. Linear theory. *J. Math. Biol.*, *36*(4), 349–388.

Doucet, A., Godsill, S., & Andrieu, C. (2000). On sequential Monte Carlo sampling methods for Bayesian filtering. *Stat. Comp.*, *10*(3), 197–208.

Doumic, M., Maia, P., & Zubelli, J. P. (2010). On the calibration of a size-structured population model from experimental data. *Acta Biotheor.*, *58*(4), 405–413.

Eden, E., Geva-Zatorsky, N., Issaeva, I., Cohen, A., Dekel, E., Danon, T., Cohen, L., Mayo, A., & Alon, U. (2011). Proteome half-life dynamics in living human cells. *Science*, *331*(6018), 764–768.

Eissing, T., Chaves, M., & Allgöwer, F. (2009). Live and let die - a systems biology view on cell death. *Comp. and Chem. Eng.*, *33*(3), 583–589.

Eissing, T., Conzelmann, H., Gilles, E., Allgöwer, F., Bullinger, E., & Scheurich, P. (2004). Bistability analyses of a caspase activation model for receptor-induced apoptosis. *J. Biol. Chem.*, *279*(35), 36892–36897.

Eissing, T., Küpfer, L., Becker, C., Block, M., Coboeken, K., Gaub, T., Goerlitz, L., Jäger, J., Loosen, R., Ludewig, B., Meyer, M., Niederalt, C., Sevestre, M., Siegmund, H.-U., Solodenko, J., Thelen, K., Telle, U., Weiss, W., Wendl, T., Willmann, S., & Lippert, J. (2011). A computational systems biology software platform for multiscale modeling and simulation: integrating whole-body physiology, disease biology, and molecular reaction networks. *Front. Physio.*, *2*, 4.

El-Samad, H. & Khammash, M. (2006). Regulated degradation is a mechanism for suppressing stochastic fluctuations in gene regulatory networks. *Biophys. J.*, *90*(10), 3749–3761.

Eldar, A. & Elowitz, M. B. (2010). Functional roles for noise in genetic circuits. *Nat.*, *467*(9), 1–7.

Elowitz, M. B., Levine, A. J., Siggia, E. D., & Swain, P. S. (2002). Stochastic gene expression in a single cell. *Science*, *297*(5584), 1183–1186.

Engl, H. W., Flamm, C., Kügler, P., Lu, J., Müller, S., & Schuster, P. (2009). Inverse problems in systems biology. *Inverse Prob.*, *25*(12), 1–51.

Evans, L. (1998). *Partial Differential Equations*. American Mathematical Society.

Fall, C. P., Marland, E. S., Wagner, J. M., & Tyson, J. J. (Eds.) (2010). *Computational cell biology*, vol. 20 of *Interdisciplinary Applied Mathematics*. Springer Berlin / Heidelberg.

Fay, M. P. & Proschan, M. A. (2010). Wilcoxon-Mann-Whitney or t-test? on assumptions for hypothesis tests and multiple interpretations of decision rules. *Stat. Surv.*, *4*, 1–39.

Feller, W. (1940). On the integro-differential equation of purely discontinous Markoff processes. *Trans. of the American Mathematical Society*, *48*, 4885–4915.

Fenton, L. F. (1960). The sum of lognormal probability distributions in scatter transmission systems. *IRE Trans. Commun. Syst.*, *8*(1), 57—67.

Fey, D. & Bullinger, E. (2009). A dissipative approach to the identification of biochemical reaction networks. In *Proc. of the 15th IFAC Symp. on Syst. Ident.*. M. Basseville, http://www.ifac-papersonline.net: IFAC-PapersOnline, vol. 15 of *Mar.*, 1259–1264.

Fox, C. (1987). *An introduction to the calculus of variations*. Dover Publ.

Fredrickson, A. G., Ramkrishna, D., & Tsuchiya, H. M. (1967). Statistics and dynamics of procaryotic cell populations. *Math. Biosci.*, *1*(3), 327–374.

Freund, R. W. (2003). Model reduction methods based on Krylov subspaces. *Acta Numerica*, *12*, 267–319.

Galán, R. F., Ermentrout, G. B., & Urban, N. N. (2007). Solving the Fokker-Planck equation with the finite-element method. *Phys. Rev. E Stat. Nonlin. Soft Matter Phys.*, *76*(5), 056110.

Gardiner, C. W. (2011). *Handbook of stochastic methods: For physics, chemistry and natural sciences*. Springer Series in Synergetics. Springer Berlin / Heidelberg, 4th ed.

Gewirtz, D. A., Holt, S. E., & Grant, S. (Eds.) (2007). *Apoptosis, senescence, and cancer*. Cancer drug discovery and development. Totowa, New Jersey: Humana Press, 2nd ed.

Geyer, C. J. & Thompson, E. A. (1995). Annealing Markov chain Monte Carlo with applications to ancestral inference. *J. Am. Stat. Assoc.*, *90*(431), 909–920.

Gillespie, D. T. (1977). Exact stochastic simulation of coupled chemical reactions. *J. Phys. Chem.*, *81*(25), 2340–2361.

Gillespie, D. T. (1992). A rigorous derivation of the chemical master equation. *Physica A*, *188*(1), 404–425.

Gillespie, D. T. (2000). The chemical Langevin equation. *J. Chem. Phys.*, *113*(1), 297–306.

Gillespie, D. T. (2001). Approximate accelerated stochastic simulation of chemically reaction systems. *J. Chem. Phys.*, *115*, 1716–1733.

Glauche, I., Herberg, M., & Roeder, I. (2010). Nanog variability and pluripotency regulation of embryonic stem cells - insights from a mathematical model analysis. *PLoS ONE*, *5*(6), e11238.

Glauche, I., Moore, K., Thielecke, L., Horn, K., Loeffler, M., & Roeder, I. (2009). Stem cell proliferation and quiescence - Two sides of the same coin. *PLoS Comput. Biol.*, *5*(7), e1000447.

Glauche, I., Thielecke, L., & Roeder, I. (2011). Cellular aging leads to functional heterogeneity of hematopoietic stem cells: a modeling perspective. *Aging Cell*, *10*, 457–465.

Golightly, A. & Wilkinson, D. J. (2010). *Markov chain Monte Carlo algorithms for SDE parameter estimation*, MIT Press, chap. 10. 253–275.

Good, P. I. (2004). *Permutation, parametric and bootstrap tests of hypotheses*. Springer Series in Statistics. Springer New York, 3 ed.

Gratzner, H. G. (1982). Monoclonal antibody to 5-bromo- and 5-iododeoxyuridine: A new reagent for detection of DNA replication. *Science*, *218*(4571), 474–475.

Green, P. J. & Mira, A. (2001). Delayed rejection in reversible jump Metropolis-Hastings. *Biometrika*, *88*(4), 1035–1053.

Grinstead, C. M. & Snell, J. L. (1997). *Introduction to probability*. American Mathematical Society.

Gutenkunst, R. N., Waterfall, J. J., Casey, F. P., Brown, K. S., Myers, C. R., & Sethna, J. P. (2007). Universally sloppy parameter sensitivities in systems biology models. *PLoS*

Comput. Biol., *3*(10), 1871–1878.

Gyllenberg, M. (1986). The size and scar distributions of the yeast Saccharomyces cervisiae. *J. Math. Biol.*, *24*(1), 81–101.

Haario, H., Laine, M., Mira, A., & Saksman, E. (2006). DRAM: Efficient adaptive MCMC. *Stat. Comp.*, *16*(4), 339–354.

Haario, H., Saksman, E., & Tamminen, J. (2001). An adaptive Metropolis algorithm. *Bernoulli*, *7*(2), 223–242.

Haasdonk, B. & Ohlberger, M. (2011). Efficient reduced models and a-posteriori error estimation for parametrized dynamical systems by offline/online decomposition. *Math. Comput. Modell. Dyn. Syst.*, *17*(2), 145–161.

Hasenauer, J., Heinrich, J., Doszczak, M., Scheurich, P., Weiskopf, D., & Allgöwer, F. (2011a). Visualization methods and support vector machines as tools for determining markers in models of heterogeneous populations: Proapoptotic signaling as a case study. In *Proc. of Workshop on Comp. Syst. Biol.*. Zürich, Switzerland, TICSP series # 57, 61–64.

Hasenauer, J., Heinrich, J., Doszczak, M., Scheurich, P., Weiskopf, D., & Allgöwer, F. (2012a). A visual analytics approach for models of heterogeneous cell populations. *EURASIP J. Bioinf. Syst. Biol.*, *4*, 10.1186/1687–4153–2012–4.

Hasenauer, J., Löhning, M., Khammash, M., & Allgöwer, F. (2012b). Dynamical optimization using reduced order models: A method to guarantee performance. *J. Process Control*, *22*(8), 1490–1501.

Hasenauer, J., Radde, N., Doszczak, M., Scheurich, P., & Allgöwer, F. (2011b). Parameter estimation for the cme from noisy binned snapshot data: Formulation as maximum likelihood problem. Extended abstract at *Conf. of Stoch. Syst. Biol.*, Monte Verita, Switzerland.

Hasenauer, J., Schittler, D., & Allgöwer, F. (2012c). Analysis and simulation of division- and label-structured population models: A new tool to analyze proliferation assays. *Bull. Math. Biol.*, *74*(11), 2692–2732.

Hasenauer, J., Waldherr, S., Doszczak, M., Radde, N., Scheurich, P., & Allgöwer, F. (2011c). Analysis of heterogeneous cell populations: a density-based modeling and identification framework. *J. Process Control*, *21*(10), 1417–1425.

Hasenauer, J., Waldherr, S., Doszczak, M., Radde, N., Scheurich, P., & Allgöwer, F. (2011d). Identification of models of heterogeneous cell populations from population snapshot data. *BMC Bioinf.*, *12*(125).

Hasenauer, J., Waldherr, S., Doszczak, M., Scheurich, P., & Allgöwer, F. (2010a). Density-based modeling and identification of biochemical networks in cell populations. In *Proc. of the 9th Int. Symp. on Dynamics and Control of Process Syst.*. Leuven, Belgium, vol. 9, 320–325.

Hasenauer, J., Waldherr, S., Radde, N., Doszczak, M., Scheurich, P., & Allgöwer, F. (2010b). A maximum likelihood estimator for parameter distributions in heterogeneous cell populations. *Procedia Computer Science*, *1*(1), 1649–1657.

Hasenauer, J., Waldherr, S., Wagner, K., & Allgöwer, F. (2010c). Parameter identification, experimental design and model falsification for biological network models using semidefinite programming. *IET Syst. Biol.*, *4*(2), 119–130.

Hasty, J., McMillen, D., Isaacs, F., & Collins, J. J. (2001). Computational studies of gene regulatory networks: in numero molecular biology. *Nat. Rev. Gen.*, *2*, 268–278.

Hasty, J., Pradines, J., Dolnik, M., & Collins, J. J. (2000). Noise-based switches and amplifiers for gene expression. *Proc. Natl. Acad. Sci. U S A*, *97*, 2075–2080.

Hawkins, E. D., Hommel, M., Turner, M. L., Battye, F. L., Markham, J. F., & Hodgkin, P. D. (2007). Measuring lymphocyte proliferation, survival and differentiation using CFSE time-series data. *Nat. Protoc.*, *2*(9), 2057–2067.

Hayflick, L. (1965). The limited in vitro lifetime of human diploid cell strains. *Exp. Cell Res.*, *37*(3), 614–636.

Hayflick, L. (1979). Progress in cytogerontology. *Mech. Ageing Dev.*, *9*(5–6), 393–408.

Hengl, S., Kreutz, C., Timmer, J., & Maiwald, T. (2007). Data-based identifiability analysis of non-linear dynamical models. *Bioinf.*, *23*(19), 2612–2618.

Henson, M. A. (2003). Dynamic modeling of microbial cell populations. *Curr. Opin. Biotechnol.*, *14*(5), 460–467.

Henson, M. A., Müller, D., & Reuss, M. (2002). Cell population modelling of yeast glycolytic oscillations. *Biochem. J.*, *368*(2), 433–446.

Herzenberg, L. A., Tung, J., Moore, W. A., Herzenberg, L. A., & Parks, D. R. (2006). Interpreting flow cytometry data: A guide for the perplexed. *Nat. Immunol.*, *7*(7), 681–685.

Hespanha, J. P. (2007). Modeling and analysis of stochastic hybrid systems. *IEE Proc. Control Theory & Applications,* Special Issue on Hybrid Systems, *153*(5), 520–535.

Higham, D. J. (2001). An algorithmic introduction to numerical simulation of stochastic differential equations. *SIAM J. Numer. Anal.*, *43*(3), 525–546.

Higham, D. J. (2008). Modeling and simulating chemical reactions. *SIAM Rev.*, *50*(2), 347–368.

Higham, D. J. & Khanin, R. (2008). Chemical master versus chemical Langevin for first-order reaction networks. *Open Appl. Math. J.*, *2*, 59–79.

Hilfinger, A. & Paulsson, J. (2011). Separating intrinsic from extrinsic fluctuations in dynamic biological systems. *Proc. Nati. Acad. Sci. U S A*, *109*(29), 12167–12172.

Hodgkin, A. L. & Huxley, A. F. (1952). A quantitative description of membrane current and its application to conduction and excitation in nerve. *J. Physiol.*, *117*(4), 500–544.

Hodgkin, P. D., Lee, J. H., & Lyons, A. B. (1996). B cell differentiation and isotype switching is related to division cycle number. *J. Exp. Med.*, *184*(1), 277–81.

Hoops, S., Sahle, S., Gauges, R., Lee, C., Pahle, J., Simus, N., Singhal, M., Xu, L., Mendes, P., & Kummer, U. (2006). COPASI – a COmplex PAthway SImulator. *Bioinf.*, *22*, 3067–3074.

Huang, S. (2010). Statistical issues in subpopulation analysis of high content imaging data. *J. Comput. Biol.*, *17*(7), 879–894.

Huh, D. & Paulsson, J. (2011). Non-genetic heterogeneity from stochastic partitioning at cell division. *Nat. Gen.*, *43*(2), 95–102.

Huys, Q. J. M. & Paninski, L. (2009). Smoothing of, and parameter estimation from, noisy biophysical recordings. *PLoS Comput. Biol.*, *5*(5), 1–16.

Hyrien, O. & Zand, M. S. (2008). A mixture model with dependent observations for the analysis of CFSE-labeling experiments. *J. American Statistical Association*, *103*(481), 222–239.

Inselberg, A. & Dimsdale, B. (1990). Parallel coordinates: a tool for visualizing multi-dimensional geometry, los alamitos, california, usa. In *Proc. of IEEE Visualization*. IEEE Computer Society Press, 361–378.

Ivanciuc, O. (2007). *Reviews in computational chemistry*, Wiley-VCH, Weinheim, vol. 23, chap. Applications of support vector machines in chemistry.

Jahnke, T. & Huisinga, W. (2007). Solving the chemical master equation for monomolecular reaction systems analytically. *J. Math. Biol.*, *54*(1), 1–26.

Janeway, C. A., Travers, P., Walport, M., & Shlomchik, M. J. (2001). *Immunobiology*. New York and London: Garland Science, 4th ed.

Jolliffe, I. T. (2002). *Principal component analysis*. Springer Series in Statistics. Springer Verlag, 2nd ed.

Joshi, M., Seidel-Morgenstern, A., & Kremling, A. (2006). Exploiting the bootstrap method for quantifying parameter confidence intervals in dynamical systems. *Metabolic Eng.*, *8*, 447–455.

Kalita, M. K., Sargsyan, K., Tian, B., Paulucci, A., Najm, H. N., Debusschere, B. J., & Brasier, A. R. (2011). Sources of cell-to-cell variability in canonical Nuclear Factor-κB (NF-κB) signaling pathway inferred from single cell dynamic images. *J. Biol. Chem.*, *286*(43), 37741–37757.

Kassem, M., Ankersen, L., Eriksen, E., Clark, B., & Rattan, S. (1997). Demonstration of cellular aging and senescence in serially passaged long-term cultures of human trabecular osteoblasts. *Osteoporosis Int.*, *7*(6), 514–524.

Kepler, T. B. & Elston, T. C. (2001). Stochasticity in transcriptional regulation: origins, consequences, and mathematical representations. *Biophys. J.*, *81*(6), 3116–3136.

Kieffer, M. & Walter, E. (2011). Guaranteed estimation of the parameters of nonlinear continuous-time models: contributions of interval analysis. *Int. J. Adapt. Control Signal Process.*, *25*(3), 191–207.

Klinke, D. J. (2009). An empirical Bayesian approach for model-based inference of cellular signaling networks. *BMC Bioinf.*, *10*(371).

Klipp, E., Herwig, R., Kowald, A., Wierling, C., & Lehrach, H. (2005a). *Systems biology in practice*. Wiley-VCH, Weinheim.

Klipp, E., Nordlander, B., Krüger, R., Gennemark, P., & Hohmann, S. (2005b). Integrative model of the response of yeast to osmotic shock. *Nat. Biotechnol.*, *23*(8), 975–982.

Kloeden, P. E. & Platen, E. (1999). *Nermical solution of stochatsic differential equations*. Berlin: Springer.

Knopp, K. (1964). *Theorie und Anwendung der unendlichen Reihen*. Springer Berlin.

Koeppl, H., Zechner, C., Ganguly, A., Pelet, S., & Peter, M. (2012). Accounting for extrinsic variability in the estimation of stochastic rate constants. *Int. J. Robust Nonlinear Control*, *22*(10), 1–21.

Komorowski, M., Costa, M. J., Rand, D. A., & Stumpf, M. P. H. (2011). Sensitivity, robustness, and identifiability in stochastic chemical kinetics models. *Proc. Natl. Acad. Sci. U S A*, *108*(21), 8645–8650.

Kramer, A., Hasenauer, J., Allgöwer, F., & Radde, N. (2010). Computation of the posterior entropy in a Bayesian framework for parameter estimation in biological networks. In *Proc. of the IEEE Multi-Conf. on Syst. and Contr.*. Yokohama, Japan, 493–498.

Kreutz, C., Bartolome Rodriguez, M. M., Maiwald, T., Seidl, M., Blum, H. E., Mohr, L., & Timmer, J. (2007). An error model for protein quantification. *Bioinf.*, *23*(20), 2747–2753.

Küpfer, L., Sauer, U., & Parrilo, P. A. (2007). Efficient classification of complete parameter regions based on semidefinite programming. *BMC Bioinf.*, *8*(12).

Lampariello, F. (2009). Ratio analysis of cumulatives for labeled cell quantification from immunofluorescence histograms derived from cells expressing low antigen levels. *Cytometry Part A*, *75A*, 665–674.

Lampariello, F. & Aiello, A. (1998). Complete mathematical modeling method for the analysis of immunofluorescence distributions composed of negative and weakly positive cells. *Cytometry*, *32*(3), 241–254.

Lang, M., Marquez-Lago, T. T., Stelling, J., & Waldherr, S. (2011). Autonomous synchronization of chemically coupled synthetic oscillators. *Bull. Math. Biol.*, *73*(11), 2678–2706.

Lee, H. Y. & Perelson, A. S. (2008). Modeling T cell proliferation and death in vitro based on labeling data: generalizations of the Smith-Martin cell cycle model. *Bull. Math. Biol.*, *70*(1), 21–44.

León, K., Faro, J., & Carneiro, J. (2004). A general mathematical framework to model generation structure in a population of asynchronously dividing cells. *J. Theor. Biol.*, *229*(4), 455–476.

Lestas, I., Vinnicombe, G., & Paulsson, J. (2010). Fundamental limits on the suppression of molecular fluctuations. *Nat.*, *467*, 174–178.

Lillacci, G. & Khammash, M. (2010a). Parameter estimation and model selection in computational biology. *PLoS Comput. Biol.*, *6*(3), e1000696.

Lillacci, G. & Khammash, M. (2010b). Parameter identification of biological networks using extended Kalman Filtering and χ^2 criteria. In *Proc. IEEE Conf. on Dec. and Contr. (CDC), Atlanta, USA*. 3367–3372.

Lipniacki, T., Paszek, P., Brasier, A. R., Luxon, B., & Kimmel, M. (2004). Mathematical model of nf-κb regulatory module. *J. Theor. Biol.*, *228*(2), 195–215.

Ljung, L. (1999). *System identification: theory for the user*. Upper Saddle River, N.J.: PTR Prentice Hall,, 2nd ed.

Löfberg, J. (2009). Dualize it: software for automatic primal and dual conversions of conic programs. *Optim. Meth. and Softw.*, *24*(3), 313–325.

Löhning, M., Hasenauer, J., & Allgöwer, F. (2011a). Steady state stability preserving nonlinear model reduction using sequential convex optimization. In *Proc. of the IEEE Conf. on Dec. and Contr.*. Orlando, Florida, USA, 7158–7163.

Löhning, M., Hasenauer, J., & Allgöwer, F. (2011b). Trajectory-based model reduction of nonlinear biochemical networks employing the observability normal form. In *Proc. of the 18th IFAC World Congress*. Milano, Italy, vol. 18, 10442–10447.

Longo, D. & Hasty, J. (2006). Dynamics of single-cell gene expression. *Mol. Syst. Biol.*, *2*(64).

Luenberger, D. G. (1968). Quasi-convex programming. *SIAM J. Appl. Math,*, *16*(5).

Luzyanina, T., Mrusek, S., Edwards, J., Roose, D., Ehl, S., & Bocharov, G. (2007a). Computational analysis of CFSE proliferation assay. *J. Math. Biol.*, *54*(1), 57–89.

Luzyanina, T., Roose, D., & Bocharov, G. (2009). Distributed parameter identification for label-structured cell population dynamics model using CFSE histogram time-series data. *J. Math. Biol.*, *59*(5), 581–603.

Luzyanina, T., Roose, D., Schenkel, T., Sester, M., Ehl, S., Meyerhans, A., & Bocharov, G. (2007b). Numerical modelling of label-structured cell population growth using CFSE distribution data. *Theor. Biol. Med. Model.*, *4*, 26.

Lyons, A. & Parish, C. (1994). Determination of lymphocyte division by flow cytometry. *J. Immunol. Methods.*, *171*(1), 131–137.

Lyons, A. B. (2000). Analysing cell division in vivo and in vitro using flow cytometric measurement of CFSE dye dilution. *J. Immunol. Methods*, *243*(1-2), 147–54.

MacKay, D. (2005). *Information theory, inference, and learning algorithms*. Cambridge University Press, 7.2 ed.

Maiwald, T. & Timmer, J. (2008). Dynamical modeling and multi-experiment fitting with potterswheel. *Bioinf.*, *24*(18), 2037–2043.

Malone, J. H. & Oliver, B. (2011). Microarrays, deep sequencing and the true measure of the transcriptome. *BMC Biol.*, *9*(34).

Mantzaris, N. V. (2007). From single-cell genetic architecture to cell population dynamics: quantitatively decomposing the effects of different population heterogeneity sources for a genetic network with positive feedback architecture. *Biophys. J.*, *92*(12), 4271–4288.

Marciniak-Czochra, A., Stiehl, T., Ho, A. D., Jäger, W., & Wagner, W. (2009). Modeling of asymmetric cell division in hematopoietic stem cells – regulation of self-renewal is essential for efficient repopulation. *Stem Cell Dev.*, *18*(3), 377–385.

Marjoram, P., Molitor, J., Plagnol, V., & Tavare, S. (2003). Markov chain Monte Carlo without likelihoods. *Proc. Nati. Acad. Sci. U S A*, *100*(26), 15324–15328.

Matera, G., Lupi, M., & Ubezio, P. (2004). Heterogeneous cell response to topotecan in a CFSE-based proliferation test. *Cytometry A*, *62*(2), 118–28.

McDonnell, K. T. & Mueller, K. (2008). Illustrative parallel coordinates. *Comput. Graphics Forum*, *27*(3), 1031–1038.

McNaught, A. D. & Wilkinson, A. (1997). *IUPAC Compendium of chemical terminology*. Blackwell Science, 2nd ed.

Meeker, W. Q. & Escobar, L. A. (1995). Teaching about approximate confidence regions based on maximum likelihood estimation. *Am. Stat.*, *49*(1), 48–53.

Merran, E., Hastings, N., & Peacock, B. (2000). *Statistical Distributions*. Wiley.

Metzger, P., Hasenauer, J., & Allgöwer, F. (2012). Modeling and analysis of division-, age-, and label-structured cell populations. In *Proc. of 9th International Workshop on Computational Systems Biology*. Ulm, Germany: Tampere International Center for Signal Processing, 60–63.

Michaelis, L. & Menten, M. (1913). Die Kinetik der Invertinwirkung. *Biochem. Z.*, *49*, 333–369.

Moles, C. G., Mendes, P., & Banga, J. R. (2003). Parameter estimation in biochemical pathways: A comparison of global optimization methods. *Genome Res.*, *13*, 2467–2474.

Müller, M. (1927). Über das Fundamentaltheorem in der Theorie der gewöhnlichen Differentialgleichungen. *Mathematische Zeitschrift*, *26*, 619–645.

Munsky, B. & Khammash, M. (2006). The finite state projection algorithm for the solution of the chemical master equation. *J. Chem. Phys.*, *124*(4), 044104.

Munsky, B. & Khammash, M. (2008). The finite state projection approach for the analysis of stochastic noise in gene networks. *IEEE Trans. Autom. Control*, *53*, 201–214.

Munsky, B. & Khammash, M. (2010). Identification from stochastic cell-to-cell variation: a genetic switch case study. *IET Syst. Biol.*, *4*(6), 356–366.

Munsky, B., Trinh, B., & Khammash, M. (2009). Listening to the noise: random fluctuations reveal gene network parameters. *Mol. Syst. Biol.*, *5*(318).

Neal, R. M. (2001). Annealed importance sampling. *Statistics and Computing*, *11*(2), 125–139.

Niepel, M., Spencer, S. L., & Sorger, P. K. (2009). Non-genetic cell-to-cell variability and the consequences for pharmacology. *Curr. Opin. Biotechnol.*, *13*(5–6), 556–561.

Nordon, R. E., Nakamura, M., Ramirez, C., & Odell, R. (1999). Analysis of growth kinetics by division tracking. *Immunol. Cell. Biol.*, *77*(6), 523–529.

Nüesch, T. (2010). *Finite state projection-based parameter estimation algorithms for stochastic chemical kinetics*. Master thesis, Swiss Federal Institute of Technology, Zürich.

Oladyshkin, S., Class, H., Helmig, R., & Nowak, W. (2011). An integrative approach to robust design and probabilistic risk assessment for CO2 storage in geological formations. *Comp. Geosciences*, *15*(3), 565–577.

Oldfield, D. G. (1966). A continuity equation for cell populations. *Bull. Math. Biol.*, *28*(4), 545–554.

Overton, W. R. (1988). Modified histogram subtraction technique for analysis of flow cytometry data. *Cytometry*, *9*(6), 619–626.

Paszek, P., Ryan, S., Ashall, L., Sillitoe, K., Harper, C. V., Spiller, D. G., Rand, D. A., & White, M. R. H. (2010). Population robustness arising from cellular heterogeneity. *Proc. Nati. Acad. Sci. U S A*, *107*(25), 1–6.

Patankar, S. V. (1980). *Numerical heat transfer and fluid flow*. Computational methods in mechanics and thermal sciences. New York: McGraw-Hill Book Company.

Paulsson, J. (2005). Models of stochastic gene expression. *Phys. of Life Rev.*, *2*(2), 157–175.

Peifer, M. & Timmer, J. (2007). Parameter estimation in ordinary differential equations for biochemical processes using the method of multiple shooting. *IET Syst. Biol.*, *1*(2), 78–88.

Pepperkok, R. & Ellenberg, J. (2006). High-throughput fluorescence microscopy for systems biology. *Nat. Rev. Mol. Cell. Biol.*, *7*(9), 690–6.

Quach, M., Brunel, N., & d'Alche Buc, F. (2007). Estimating parameters and hidden variables in non-linear state-space models based on odes for biological networks inference. *Bioinf.*, *23*(23), 3209–3216.

Rao, C. V. & Arkin, A. P. (2003). Stochastic chemical kinetics and the quasi-steady-state assumption: Application to the gillespie algorithm. *J. Chem. Phys.*, *118*(11), 4999–5010.

Rathinam, M., Petzold, L., Cao, Y., & Gillespie, D. T. (2003). Stiffness in stochastic chemically reaction systems: The implicit tau-leaping method. *J. Chem. Phys.*, *119*(24), 12784–12794.

Raue, A., Kreutz, C., Maiwald, T., Bachmann, J., Schilling, M., Klingmüller, U., & Timmer, J. (2009). Structural and practical identifiability analysis of partially observed dynamical models by exploiting the profile likelihood. *Bioinf.*, *25*(25), 1923–1929.

Raue, A., Kreutz, C., Maiwald, T., Klingmüller, U., & Timmer, J. (2011). Addressing parameter identifiability by model-based experimentation. *IET. Syst. Biol.*, *5*(2), 120–130.

Reinker, S., Altman, R. M., & Timmer, J. (2006). Parameter estimation in stochastic biochemical reactions. *IEE Proc. Syst. Biol.*, *153*(4), 168–178.

Renart, J., Reiser, J., & Stark, G. R. (1979). Transfer of proteins from gels to diazobenzyloxymethyl-paper and detection with antisera: a method for studying antibody specificity and antigen structure. *Proc. Nati. Acad. Sci. U S A*, *76*(7), 3116–3120.

Reshef, D. N., Reshef, Y. A., Finucane, H. K., Grossman, S. R., McVean, G., Turnbaugh, P. J., Lander, E. S., Mitzenmacher, M., & Sabeti, P. C. (2011). Detecting novel associations in large data sets. *Science*, *334*(6062), 1518–1524.

Revy, P., Sospedra, M., Barbour, B., & Trautmann, A. (2001). Functional antigen-independent synapses formed between T cells and dendritic cells. *Nat. Immunol.*, *2*(10), 925–931.

Risken, H. (1996). *The Fokker-Planck equation: Methods of solution and applications*. Springer Berlin / Heidelberg, 2nd ed.

Rodgers, J. L. & Nicewander, W. A. (1988). Thirteen ways to look at the correlation coefficient. *Am. Stat.*, *42*(1), 59–66.

Rosenfeld, N., Young, J. W., Alon, U., Swain, P. S., & Elowitz, M. B. (2005). Gene regulation at the single-cell level. *Science*, *307*(5717), 1962–1965.

Ruess, J., Milias, A., Summers, S., & Lygeros, J. (2011). Moment estimation for chemically reacting systems by extended Kalman filtering. *J. Chem. Phys.*, *135*(165102).

Rumschinski, P., Borchers, S., Bosio, S., Weismantel, R., & Findeisen, R. (2010). Set-based dynamical parameter estimation and model invalidation for biochemical reaction networks. *BMC Syst. Biol.*, *4*(69).

Runge, C. (1901). über empirische Funktionen und die Interpolation zwischen äquidistanten Ordinaten. *Zeitschrift für Mathematik und Physik*, *46*, 224–243.

Schittler, D., Hasenauer, J., & Allgöwer, F. (2011). A generalized population model for cell proliferation: Integrating division numbers and label dynamics. In *Proc. of 8th International Workshop on Computational Systems Biology*. Zürich, Switzerland: Tampere International Center for Signal Processing, TICSP series # 57, 165–168.

Schittler, D., Hasenauer, J., Allgöwer, F., & Waldherr, S. (2010). Cell differentiation modeled via a coupled two-switch regulatory network. *Chaos*, *20*(4), 045121.

Schlatter, R., Schmich, K., Vizcarra, I. A., Scheurich, P., Sauter, T., Borner, C., Ederer, M., Merfort, I., & Sawodny, O. (2009). ON/OFF and beyond – a boolean model of apoptosis. *PLoS Comput. Biol.*, *5*(12), 1–13.

Schmidt, H. & Jirstrand, M. (2006). Systems biology toolbox for MATLAB: a computational platform for research in systems biology. *Bioinf.*, *22*(4), 514–515.

Schöberl, B., Eichler-Jonsson, C., Gilles, E. D., & Muller, G. (2002). Computational modeling of the dynamics of the MAP kinase cascade activated by surface and internalized EGF receptors. *Nat. Biotechnol.*, *20*, 370–375.

Schöberl, B., Pace, E. A., Fitzgerald, J. B., Harms, B. D., Xu, L., Nie, L., Linggi, B., Kalra, A., Paragas, V., Bukhalid, R., Grantcharova, V., Kohli, N., West, K. A., Leszczyniecka, M., Feldhaus, M. J., Kudla, A. J., & Nielsen, U. B. (2009). Therapeutically targeting ErbB3: A key node in ligand-induced activation of the ErbB receptor–PI3K axis. *Science Signaling*,

2(77), ra31.

Schölkopf, B., Sung, K. K., Burges, C. J. C., Girosi, F., Niyogi, P., Poggio, T., & Vapnik, V. (1997). Comparing support vector machines with Gaussian kernels to radial basis function classifiers. *IEEE Trans. Signal Process.*, *45*, 2758–2765.

Schroeder, T. (2011). Long-term single-cell imaging of mammalian stem cells. *Nat. Methods*, *8*(4), 30–35.

Schwarz, G. (1978). Estimating the dimension of a model. *Ann. Statist.*, *6*(2), 461–464.

Scott, D. W. (1992). *Multivariate density estimation: theory, practice, and visualization*. New York, Chichester: John Wiley & Sons.

Shcheprova, Z., Baldi, S., Frei, S. B., Gonnet, G., & Barral, Y. (2008). A mechanism for asymmetric segregation of age during yeast budding. *Nat.*, *454*(7205), 728–734.

Silverman, B. W. (1986). *Density estimation for statistics and data analysis*. Monographs on Statistics and Applied Probability. London: Chapman and Hall.

Singh, A. & Hespanha, J. P. (2011). Approximate moment dynamics for chemically reacting systems. *IEEE Trans. Autom. Control*, *56*(2), 414–418.

Singh, D. K., Ku, C.-J., Wichaidit, C., Steininger, R. J., Wu, L. F., & Altschuler, S. J. (2010). Patterns of basal signaling heterogeneity can distinguish cellular populations with different drug sensitivities. *Mol. Syst. Biol.*, *6*(369).

Sinko, J. & Streifer, W. (1967). A new model for age-size structure of a population. *Ecology*, *48*(6), 910–918.

Sisson, S. A., Fan, Y., & Tanaka, M. M. (2007). Sequential Monte Carlo without likelihoods. *Proc. Natl. Acad. Sci. U S A*, *104*(6), 1760–1765.

Smith, J. A. & Martin, L. (1973). Do cells cycle? *Proc. Nati. Acad. Sci. U S A*, *70*(4), 1263–1267.

Smola, A. J. & Schölkopf, B. (2004). A tutorial on support vector regression. *Stat. Comp.*, *14*(3), 199–222.

Snijder, B. & Pelkmans, L. (2011). Origins of regulated cell-to-cell variability. *Nat. Rev. Mol. Cell. Biol.*, *12*(2), 119–125.

Song, C., Phenix, H., Abedi, V., Scott, M., Ingalls, B. P., Kærn, M., & Perkins, T. J. (2010). Estimating the stochastic bifurcation structure of cellular networks. *PLoS Comput. Biol.*, *6*(3), e1000699.

Spencer, S. L., Gaudet, S., Albeck, J. G., Burke, J. M., & Sorger, P. K. (2009). Non-genetic origins of cell-to-cell variability in TRAIL-induced apoptosis. *Nat.*, *459*(7245), 428–433.

Spencer, S. L. & Sorger, P. K. (2011). Measuring and modeling apoptosis in single cells. *Cell*, *144*(6), 926–939.

Stamatakis, M. & Zygourakis, K. (2010). A mathematical and computational approach for integrating the major sources of cell population heterogeneity. *J. Theor. Biol.*, *266*(1), 41–61.

Stiehl, T. & Marciniak-Czochra, A. (2011). Characterization of stem cells using mathematical models of multistage cell lineages. *Math. Comp. Modelling*, *53*(7–8), 1505–1517.

Stone, C. J. (1984). An asymptotically optimal window selection rule for kernel density estimation. *Annu. Stat.*, *12*(4), 1285–1297.

Strang, G. & Fix, G. (1973). *An analysis of the finite element method*. Prentice Hall.

Surulescu, C. & Surulescu, N. (2010). A nonparametric approach to cells dispersal. *Int. J. Biomath. Biostat.*, *1*, 109–128.

Swain, P. S., Elowitz, M. B., & Siggia, E. D. (2002). Intrinsic and extrinsic contributions to stochasticity in gene expression. *Proc. Natl. Acad. Sci. U S A*, *99*(20), 12795–12800.

Tay, S., Hughey, J. J., Lee, T. K., Lipniacki, T., Quake, S. R., & Covert, M. W. (2010). Single-cell nf-κb dynamics reveal digital activation and analogue information processing. *Nature*, *466*, 267–271.

Thompson, W. C. (2012). *Partial differential equation modeling of flow cytometry data from CFSE-based proliferation assays*. Ph.d. thesis, North Carolina State University.

Toni, T. & Stumpf, M. P. H. (2010). Simulation-based model selection for dynamical systems in systems and population biology. *Bioinf.*, *26*(1), 104–110.

Toni, T., Welch, D., Strelkowa, N., Ipsen, A., & Stumpf, M. P. H. (2009). Approximate Bayesian computation scheme for parameter inference and model selection in dynamical systems. *J. R. Soc. Interface*, *6*, 187–202.

Trucco, E. (1965). Mathematical models for cellular systems the von foerster equation. Part i. *Bull. Math. Biol.*, *27*(3), 285–304.

Tsien, R. (1998). The green fluorescent protein. *Annu. Rev. Biochem.*, *67*, 509–544.

Tsuchiya, H. M., Fredrickson, A. G., & Aris, R. (1966). Dynamics of microbial cell populations. *Adv. Chem. Eng.*, *6*, 125–206.

Turing, A. M. (1952). The chemical basis of morphogenesis. *Phil. Trans. Royal Soc. London Series B*, *237*(641), 37–72.

Turlach, B. A. (1993). Bandwidth selection in kernel density estimation: A review. Tech. rep., Institut de Statistique, Voie du Roman Pays 34, B-1348 Louvain-la-Neuve, Belgium.

van Kampen, N. G. (2007). *Stochastic processes in physics and chemistry*. Amsterdam: North-Holland, 3rd revised edition ed.

Vaz, A. & Vicente, L. (2007). A particle swarm pattern search method for bound constrained global optimization. *J. Global Optim.*, *39*(2), 197–219.

Villadsen, J. & Michelsen, M. L. (1978). *Solution of differential equation models by polynomial approximation*. Prentice-Hall.

von Foerster, H. (1959). Some remarks on changing populations. *The kinetics of cellular proliferation*, New York: Grune and Stratton. 382–407.

Vyshemirsky, V. & Girolami, M. A. (2008). Bayesian ranking of biochemical system models. *Bioinf.*, *24*(6), 833–839.

Wajant, H., Pfizenmaier, K., & Scheurich, P. (2003). Tumor necrosis factor signaling. *Cell Death Differ.*, *10*(1), 45–65.

Waldherr, S. (2009). *Uncertainty and robustness analysis of biochemical reaction networks via convex optimisation and robust control theory*. No. 1163 in 705 Fortschrittsberichte VDI Reihe 8. Düsseldorf, Germany: VDI Verlag.

Waldherr, S., Hasenauer, J., & Allgöwer, F. (2009). Estimation of biochemical network parameter distributions in cell populations. In *Proc. of the 15th IFAC Symp. on Syst. Ident.*. Saint-Malo, France, vol. 15, 1265–1270.

Waldherr, S., Wu, J., & Allgöwer, F. (2010). Bridging time scales in cellular decision making with a stochastic bistable switch. *BMC Syst. Biol.*, *4*(108).

Walker, D. C., Georgopoulos, N. T., & Southgate, J. (2008). From pathway to population: a multiscale model of juxtacrine EGFR-MAPK signalling. *BMC Syst. Biol.*, *2*(102).

Wang, H. & Huang, S. (2007). Mixture-model classification in DNA contant analysis. *Cytometry Part A*, *71*(9), 716–723.

Watson, J. V. (2001). Proof without prejudice revisited: immunofluorescence histogram analysis using cumulative frequency subtraction plus ratio analysis of means. *Cytometry*, *43*(1), 55–68.

Weber, P., Hasenauer, J., Allgöwer, F., & Radde, N. (2011). Parameter estimation and identifiability of biological networks using relative data. In *Proc. of the 18th IFAC World Congress*. Milano, Italy, vol. 18, 11648–11653.

Weise, T. (2009). Global optimization algorithms: Theory and application. ebook, Nature Inspired Computation and Applications Laboratory (NICAL), University of Science and Technology, China.

Weiße, A. Y., Middleton, R. H., & Huisinga, W. (2010). Quantifying uncertainty, variability and likelihood for ordinary differential equation models. *BMC Syst. Biol.*, *4*(144).

Welch, B. L. (1947). The generalization of "student's" problem when several different population variances are involved. *Biometrika*, *34*(1–2), 28–35.

Wilkinson, D. J. (2007). Bayesian methods in bioinformatics and computational systems biology. *Briefings in Bioinf.*, *8*(2), 109–116.

Wilkinson, D. J. (2009). Stochastic modelling for quantitative description of heterogeneous biological systems. *Nat. Rev. Genet.*, *10*(2), 122–133.

Wilkinson, D. J. (2010). Parameter inference for stochastic kinetic models of bacterial gene regulation: A Bayesian approach to systems biology. In *Proc. of 9th Valencia Int. Meet. (Bayesian Statistics 9), Valencia, Spain*. Oxford University Press, 679–705.

Witt, J., Barisic, S., Schumann, E., Allgöwer, F., Sawodny, O., Sauter, T., & Kulms, D. (2009). Mechanism of pp2a-mediated IKKβ dephosphorylation: a systems biological approach. *BMC Syst. Biol.*, *3*(71).

Yang, R., Niepel, M., Mitchison, T. K., & Sorger, P. K. (2010). Dissecting variability in responses to cancer chemotherapy through systems pharmacology. *Clin. Pharmacol. Ther.*, *88*(1), 34–38.

Yates, A., Chan, C., Strid, J., Moon, S., Callard, R., George, S. J. T., & Stark, J. (2007). Reconstruction of cell population dynamics using CFSE. *BMC Bioinf.*, *8*(196).

Zechner, C., Ruess, J., Krenn, P., Pelet, S., Peter, M., Lygeros, J., & Koeppl, H. (2012). Moment-based inference predicts bimodality in transient gene expression. *Proc. Nati. Acad. Sci. U S A*, *109*(21), 8340–8345.

Zwietering, M. H., Jongenburger, I., Rombouts, F. M., & van 't Riet, K. (1990). Modeling of the bacterial growth curve. *Appl. Environ. Microbiol.*, *56*(6), 1875–1881.